STATISTICS FOR
SOCIAL WORKERS

Third Edition

STATISTICS FOR SOCIAL WORKERS

Robert W. Weinbach
The University of South Carolina

Richard M. Grinnell, Jr.
The University of Calgary

Statistics for Social Workers, Third Edition

Longman, 10 Bank Street, White Plains, N.Y. 10606

Associated companies:
Longman Group Ltd., London
Longman Cheshire Pty., Melbourne
Longman Paul Pty., Auckland
Copp Clark Longman Ltd., Toronto

Senior acquisitions editor: David Shapiro
Development editor: Susan G. Alkana
Production editor: ProBook/Linda Moser
Cover design: Renée Kilbride Edelman
Production supervisor: Richard Bretan

Library of Congress Cataloging-in-Publication Data

Weinbach, Robert W.
    Statistics for social workers / Robert W. Weinbach, Richard M.
Grinnell, Jr.—3rd ed.
        p.   cm.
    Includes bibliographical references and index.
    ISBN 0-8013-1388-0
    1. Social sciences—Statistical methods.   2. Social service—
Statistical methods.   3. Statistics.   I. Grinnell, Richard M.
II.   Title.
HA29.W433   1994
519.5′024362—dc20   94-6160
                                                            CIP

2 3 4 5 6 7 8 9 10-CRS-99 98 97 96 95

Contents

v

CHAPTER 6 **SELECTING STATISTICAL TESTS**   93

CHAPTER 7 **CROSS-TABULATION**   107

Preface

The favorable reception afforded the preceding two editions of *Statistics for Social Workers* has encouraged us to write a third. As before, we continue to believe that social work students and practitioners have a wide range of mathematical backgrounds and varying attitudes toward statistics. However, they tend to share one common characteristic—they are "people oriented." They have selected their profession because they want to work with people, not numbers or abstract concepts. We believe, nonetheless, that if they can see how numbers and statistical analysis can be used to help them to become more effective practitioners, they will approach statistics in a positive manner.

AUDIENCE

This book is intended for social work students (undergraduate or graduate) and practitioners who are either new to statistical analysis or may have encountered it before but never really understood it. The book assumes no prior knowledge of the topic. We have presented the material in the most unintimidating way possible, using a crisp, readable style and simple language, while avoiding difficult words and convoluted phrases. The text is easy to teach from and to learn from. Figures and case examples have been used extensively throughout to provide visual representations of the statistical concepts presented.

As with the previous two editions, we continue to stress our belief that statistics is relevant to the development and refinement of those helping skills that social work practitioners require. Students who read this book will soon discover that statistical procedures are powerful tools that can help them answer questions about social work practice and can help them to become more effective practitioners.

GOAL

The goal of this book is to meet the challenge of introducing today's people-oriented social work students and practitioners to a topic that they might not choose to study. It seeks to explain statistical analysis while keeping in mind social workers' diverse mathematical backgrounds and interests. Consequently, it is as "unmathematical" as we dared make it, while still providing enough mathematical explanation to present the indispensable statistical concepts. It contains the language and concepts needed for beginning statistical literacy.

We hope that our modest goal will permit students to make decisions about when statistical analysis is necessary, to evaluate theoretical or practice-based social work problems, to determine which type of analysis is most appropriate to use in a given research situation, to interpret statistical results, and to communicate the statistical results and interpretations in terms that are meaningful for others. In short, this text is designed to help social workers to appreciate, interpret, use, and integrate statistics within their professional practice. Other, more advanced texts and courses are required for them to become statisticians.

RELIANCE ON COMPUTERS

You have probably heard the slogan "let your fingers do the walking." Well, our slogan is "let the computer do the work." We have avoided the inclusion of unnecessary formulas and their derivations while focusing instead on what a statistic really means. Why have we done this? The prices of personal computers and statistical software have continued to decline to the point where most students and practitioners now have access to them. Many schools and departments of social work, as well as most social agencies, have computer equipment that allows social workers to perform word processing as well as data entry and analysis.

The current state of computer technology suggests an analogy to taking a trip on an airplane. Do we have to know the laws of aerodynamics to fly in an airplane? Or do we just have to know how to make a reservation and hope that the people who built and fly the plane know those laws well? Does a social work researcher or practitioner need to know how to perform all the mathematical calculations of a statistical test just to use it correctly? Or does he or she just have to know when to use the test and what to tell the computer to have it perform the correct analysis on a given data set?

This book will help the social work student to choose the correct statistical analysis, although it will offer little to help him or her do the calculations based on statistical probability theory—we now leave that up to the computer. We believe that social work students and practitioners have better things to do with their time than trying to compute a statistic mathematically when a computer can do the job in seconds. The chances of a social worker in the 1990s doing a statistical analysis by hand with a large data set are small. We would rather have him or her spend valuable time determining what a particular statistic means and how to use it, rather than trying to determine the meaning of a formula and plugging in an endless series of numbers to get a statistical outcome that he or she may not understand.

We are aware of the limitations of not providing formulas for most statistics in this book and of not showing how the formulas are used in their calculations. There is a great deal of truth to the statement that students may not totally understand and appreciate a statistic unless they are shown how its formula is derived. To return to our earlier analogy, we admit readily that an airplane passenger who knows the laws of aerodynamics will unquestionably have a greater understanding and appreciation of the achievement of flight than a passenger who does not. However, they will accomplish their mutual goal: Both will arrive at the same destination at the same time.

Very few people would fly if they were required to know the laws of aerodynamics before they could take a trip. In the same vein, very few social workers would use statistical analyses if they were required to know first the laws of probability and statistical theory. Many people appreciate and use airplanes without knowing the laws of aerodynamics, and social workers can be taught to appreciate and use statistics without knowing the laws of probability theory or of the derivation formulas. However, if a reader of this book wishes to further develop his or her statistical literacy by studying the formulas upon which statistical analyses are based, there are many more advanced books and courses that can assist in this undertaking.

CONTENTS

The contents of this book have been selected and arranged so that the book can be used as a supplementary statistics text for a social work research methods course or as the primary text in an introductory social work statistics course. The book begins by introducing basic concepts and definitions that are necessary to understand statistical analysis. These concepts are presented in Chapter 1 and include such terms as data, variables, constants, conceptualization, operationalization, reliability, validity, and levels of measurement.

Commonly used descriptive statistics are covered in Chapters 2-4. These three chapters familiarize the student with basic statistical language and procedures through an examination of the most basic form of statistical analysis—that which is designed to summarize and display the distribution of a single variable.

Chapters 5–6 introduce the reader to the related topics of inferential statistics and hypothesis testing. The remaining six chapters (Chapters 7–12) present a wide array of statistical methods commonly used by social workers to test hypotheses about relationships between and among variables at different levels of measurement.

We have used numerous micro and macro social work practice examples to illustrate the use of the basic statistical concepts presented. Through these examples, we have tried to demonstrate that statistics can help us answer every-day social work questions. The answers, in turn, can be used to enhance our effectiveness as practitioners.

CHANGES IN THIS EDITION

In writing this edition, we responded to many of the professors and students who generously critiqued the two earlier editions. We sought to retain what worked and to revise those areas that needed changes. While the book has kept its identity, it also is much more comprehensive than the previous two editions.

A quick glance at the Contents reveals that the book has undergone some major as well as minor changes. The most obvious changes are the addition of a new chapter on simple regression and coverage of analysis of variance. Another new chapter, an overview of the most commonly used types of multivariate analysis, was also included as a response to reviewers who requested additional statistical coverage. The entire book has been enhanced with new and updated examples, tables, and figures. To make room for all the new content, some topics (those that were judged to require less extensive discussion) were condensed and others combined.

ACKNOWLEDGMENTS

We would like to acknowledge our academic deans, Frank B. Raymond, III, of the College of Social Work, The University of South Carolina, and Ray J. Thomlison of the Faculty of Social Work, The University of Calgary, who provided us with encouragement and support. We also express our deep appreciation to Marj Andrukow, for her conscientious and skillful preparation of the tables and figures, and to Denise Ashurst, for her assistance in the preparation of the final manuscript.

Many people have contributed to the completion of this book. Hundreds of students, through questions and challenges posed during class sessions over the past 20 years, have served to shape the authors' thoughts about how statistical analysis can be taught in a meaningful and "user friendly" manner. A number of faculty colleagues have kindly shared their expertise by commenting on earlier drafts of the text or by commenting on coverage in the earlier editions.

The following people may take partial credit for whatever achievements and improvements this edition represents; we alone must accept responsibility for its shortcomings:

Hailv Abatena, University of Nevada—Las Vegas

William E. Berg, University of Wisconsin—Milwaukee

Peter Gabor, The University of Calgary

John T. Gandy, The University of South Carolina

Leon H. Ginsberg, The University of South Carolina

Arnold L. Greenfield, Michigan State University

Bud Hanson, University of Windsor

Keith M. Kilty, Ohio State University

Judy Krysik, The University of Calgary

Grant McDonald, York University

Steven L. McMurtry, Arizona State University

Cathy Pike, The University of South Carolina

Paul R. Raffoul, University of Houston

Michael Rothery, The University of Calgary

Allen Rubin, The University of Texas at Austin

Michael J. Sheridan, Virginia Commonwealth University

Jackie Sieppert, The University of Calgary

Barbara Thomlison, The University of Calgary

Leslie M. Tutty, The University of Calgary

Yvonne Unrau, The University of Calgary

Alex Westerfelt, University of Kansas

G. Robert Whitcomb, The University of South Carolina

If the material presented in this book helps social work students and practitioners to develop or expand their statistical knowledge base and assists them in preparing for more advanced statistics courses, our efforts will have been more than justified. If it also encourages them to appreciate the place of statistics in professional social work practice, our task will have been fully rewarded.

chapter 1

Introduction

Effective professional social workers require an understanding of statistical analysis. A knowledge of statistics is crucial to many decisions that affect their ability to be successful in their practices and to communicate knowledge to others.

The National Association of Social Workers and the Council on Social Work Education strongly encourage social workers to develop their ability to understand the results of statistical analyses contained in social work journal articles and books. This requires statistical literacy—a knowledge of the conventional methods used to gather, sort, organize, and analyze data. Unless we know whether a statistical analysis was performed correctly, we cannot know whether the findings of a research study have credibility or to what degree recommendations derived from the analysis should influence our work with clients.

Responsible social workers also seek to contribute to the profession's knowledge base by publishing the results of their own research studies. This involves communicating the knowledge acquired through their research studies and practice experiences to others who can use it. If the knowledge that we seek to communicate is to be credible, we must demonstrate that our data were generated, analyzed, and interpreted according to established procedures. Fortunately, as we shall see, these procedures are based on logic that social workers already use in many practice situations. We apply them in a similar manner whether we collect and analyze data for a "practice" situation or for a "research" study.

In addition, as responsible social work professionals, we need to evaluate regularly our own practice effectiveness. To do this, we must rely on more than personal insight, intuition, or the feeling that we are (or are not) being effective; we must also employ objective evaluative methods. Statistical methods do not guarantee objectivity, but they have the potential to enhance it. Understanding and using statistical procedures appropriately allow social workers to move toward the objective of empirically based practice.

BASIC CONCEPTS

Sound statistical analyses are dependent upon the measurement process used to generate and analyze data. Statistical analyses produce meaningful results *only* when sound methods for data collection have been employed. Even the best statistical test is useless if the data examined are of questionable value. We will begin by reviewing key terms and concepts that relate to the nature of data that are collected and analyzed statistically.

Data

The words *data* and *information* are sometimes used interchangeably in statistics textbooks. However, in this book, a distinction is made between them. *Data* will refer to numerical results obtained by a research study *before analysis:* for example, the score obtained when a self-report standardized measuring instrument is completed by a client. *Information* will refer to the interpretation of these results *after analysis:* for example, the finding that a certain treatment intervention was successful in significantly reducing substance abuse in 52 percent of the clients studied; or an estimate that only 20 percent of the cases of sexual abuse involving people who are physically disabled in the United States are ever reported. In other words, data are collected and analyzed in a research study in order to produce information. A sufficient amount of information, properly collated with other information, leads to knowledge building.

Practitioners and researchers engage in many types of data gathering. Our profession uses a wide variety of research methods to collect data about various aspects of our profession and about the clients we serve. Social workers use surveys, interviews, content analyses, experiments, and direct observation of behavior in natural settings. In addition, we use existing data already observed and recorded for other purposes, such as police, agency, and hospital records, and census materials.

Data are the starting point for statistical analyses; therefore, they must be highly accurate when collected. No conclusions based on a research study's findings can be any better than the data that contributed to those conclusions. Statistical analyses of inaccurate or misleading data are worse than no analysis at all—they can lead to wrong conclusions and to recommendations that can negatively affect the work of the social work practitioner.

Variables and Constants

There is always a limit to how much data a researcher can collect and use in any given research study. Like a social worker conducting an intake interview, a social work researcher does not need to know everything about his or her research participants or objects. Both practitioners and researchers limit their data collection to only those data that will be used for decision making. In research studies we call those areas of interest variables.

Variables are characteristics that vary among the research participants (or objects) studied. Among human beings, examples of variables might be educational levels, genders, sexual orientations, ethnicities, motivational levels, stress levels, self-esteem levels, and so on. In contrast to variables, traits or characteristics that are common to all research participants (or objects) are known as *constants*. An example of a constant among all human beings is mortality. Researchers are primarily interested in variables. They want to learn why variations occur among participants (or objects) and what other variables may relate to these variations. For example, a researcher may study a sample of crack cocaine–addicted prisoners in an attempt to learn what might explain their patterns of drug abuse. Crack addiction would be a constant, and their patterns of drug abuse would be a variable.

Values of Variables. The different measurements that a variable can take are referred to as *values*. For example, the variable *gender* can be measured by classifying clients using only two values—male and female; we can measure the variable *age* by using its value at the person's last birthday. The variable *number of active clients in different social work agencies in a given month* also would have different values—the different client census counts that occur among the agencies. The variable *highest social work degree attained* might be broken down into four values (e.g., 1, 2, 3, and 4) based on responses to this questionnaire item:

What is your highest social work degree? (Circle one number below.)

1. Associate
2. B.S.W.
3. M.S.W.
4. Ph.D./D.S.W.

Frequencies of Values. More often than not, a given value occurs more than once in a group of research participants (or objects). The number of times that it occurs is referred to as its *frequency*. Frequencies for values of a variable usually vary. For example, within a sample of agency clients, the frequency for the value *African American* might be 16 for the variable *race,* while the frequency for the value *Asian American* might be 12. Or the number of social workers in an agency broken down by their highest social work degree received might be 4 associate degrees, 12 B.S.W.s, 19 M.S.W.s, and 2 Ph.D./D.S.W.s.

CONCEPTUALIZATION

Often when a research study is begun, the research problem appears complex. It may seem to take many forms and to be related, or possibly related, to many other phenomena that in themselves may be difficult to understand. In social work practice and research we seek to make this tangle manageable by selecting and identifying what we feel to be the most important or relevant variables and stating the values that each variable can assume. At the same time, we can also identify possible interconnections between the variables.

Selecting the most relevant variables to examine is called *conceptualization*. It also involves specifying as precisely as possible what we have in mind when we refer to those variables. For example, the prediction that "Among children from ages 3 to 10 Intervention A will decrease autistic behavior more than Intervention B" implies a conviction that the two different interventions (A and B) affect autistic behavior differently.

In order to examine the relationship between types of interventions utilized and their respective abilities to decrease the autistic behavior of these children, we must clearly state the meaning of both variables (i.e., intervention and autistic behavior) and the proposed relationship between them. For instance, the conceptualization of exactly what constitutes Intervention A might include the introduction of shaping techniques once a week, the application of positive reinforcement during school hours, or both. Autistic behavior can be conceptualized to include self-destructive behaviors during school hours, responses on a measuring instrument that measures the severity of autism completed by school personnel, social isolation, or all three.

Without first delineating the meaning of these two variables, it would be useless to attempt to establish a relationship between them. Why would we want to see if there was a relationship between two variables if we did not know what the real meanings of the variables were in the first place? This is why conceptualization is so important.

OPERATIONALIZATION

Specifying the measuring devices that will be used to measure the variables that we have conceptualized is referred to as *operationalization*. It further clarifies the meaning of variables by reducing them to measurable dimensions. Operationalization attaches either qualitative (differences in the type or kind of a variable) or quantitative (differences in the amount of a variable) meanings to the different values of the variables.

The specific methods used to measure variables will differ from place to place, time to time, person to person, and study to study. For example, in different research situations, measurement of the variable *self-esteem* might include observations of client behaviors, subjective diagnoses of clients by their social workers, self-reports of clients, or client responses on self-administered standardized measuring instruments.

Often we measure variables that cannot be seen directly. Various methods have been developed for use in these situations. One method is to record the verbal, written, and/or physical responses of research participants to specific stimuli. For example, we may wish to find out the extent to which computerization in a particular social work agency has affected the level of job satisfaction of the social workers employed there. The job satisfaction of the social workers could be measured before and after the computerization. Methods of measurement might include asking workers directly or having them complete a self-report job satisfaction questionnaire. We might also measure a behavioral response such as job absenteeism, which previous research studies have indicated is directly related to job satisfaction. If at all possible, it would be desirable to use all three of these indicators to contribute to our overall measurement of job satisfaction.

Selecting specific meaningful variables to study (conceptualization) and finding sound ways to measure them (operationalization) are important phases in the social work research process. If, for example, we want to learn something about the relative effectiveness of two different interventions (e.g., individual treatment versus group treatment) for treating depressed clients, we must provide a clear conceptualization and operationalization of both the two interventions and of depression.

. One indicator of clients' depression levels might be records of clients' appearance made by the social worker, who evaluates how depressed they appear to be. Used alone to measure clients' depression levels, such a method of measurement may not yield accurate results. Another indicator that could be used to measure depression might be each client's score on a self-report standardized measurement instrument that measures depression. We would hope that all client scores viewed together might suggest a pattern. It might appear, for example, that clients who were treated individually scored lower on the measuring instrument than clients who were treated in a group. However, we would be premature if we concluded, on the basis of the apparent pattern of scores, that treatment effectiveness (reduction of depression) and treatment methods (individual or group) are, in fact, really related.

Statistical analyses can take some of the guesswork out of any conclusions we wish to make regarding the relationship between two or more variables. The results of statistical testing can help us make more informed statements about the treatment effectiveness variable and how it may (or may not) be related to depression levels. Whether or not the results of statistical testing ultimately suggest to us that individual treatment is more effective than group treatment depends largely upon our assessment of how well the self-report standardized measuring instrument measured the variable *depression*. Statistical tests only perform mathematical operations on data; they have no way of knowing whether or not the measurement that generated the data was reliable and valid.

EVALUATING DATA

In research, certain criteria are used to judge the quality of data. For a detailed discussion of these criteria, the reader will need to consult a social work research methods text. We will only briefly discuss the two criteria that are most commonly employed to evaluate the quality of measurement that produces data: reliability and validity.

Reliability

In the simplest of terms, *reliability* refers to the consistency of a measurement. It answers the question, "To what degree does the measurement produce similar results in measuring people (or objects) under similar conditions?" One measurement that proposes to measure a variable (such as depression) may produce very consistent measurements. Therefore, it is considered reliable. Another measurement of depression may be influenced by such factors as who is performing the measuring, the time of day or year that the measurement took place, and so on. Thus, it would be considered less reliable than the first measurement.

Validity

Even perfect reliability (which rarely exists) in itself does not guarantee that the method or measuring instrument used to measure a variable will produce the desired results. It can be very consistent in producing the same measurements, but the measurements can be consistently wrong! How can this be? The method or measuring instrument may be biased or distorted in some way. For example, a measurement believed to measure depression may produce consistent results. However, if the measurement really is measuring something other than depression (such as self-esteem), it is not producing an accurate measurement of depression and is not an appropriate means of measurement.

A cloth tape measure may be perfectly appropriate for measuring the width of various tables, but if it has shrunk from being left in the rain, it will not produce valid measurements. It will produce very consistent (reliable) results under a variety of measurement conditions, but all measurements that it produces will suggest that tables measured with it are longer than they really are. In other words, they will not be accurate.

However, if a measurement of a variable is both reliable (consistent) and unbiased, it will produce accurate results. We then describe the measurement as valid. The conclusion that a measurement has *validity* is a conclusion that it truly measures what it is supposed to measure *and* that it measures it accurately. Valid measurement of variables produces data that, when statistically analyzed correctly, can generate valuable knowledge.

Validity and reliability are clearly related. If an instrument measures what it is supposed to measure, it is valid, and, by definition, it must also be reliable. There cannot be validity without reliability, but there can be reliability without validity: High reliability does not guarantee validity. Reliability indicates that a consistent measurement is occurring, but what is being measured may or may not be the variable that a researcher is attempting to measure.

HYPOTHESES

As will be discussed in Chapter 5, many statistical analyses are devoted to the task of testing hypotheses. One common definition of a *hypothesis* is that it is a statement of a relationship between two or more variables. In traditional quantitative forms of research, a hypothesis expresses what, based primarily on an extensive review of the professional literature, we believe to be true. A hypothesis is stated in such a way that it can gain support (or not gain support) through statistical analyses. For example, we might state a hypothesis about the two variables *depression* and *sleep patterns* as follows:

> **Hypothesis 1:** People who are depressed will have different sleep patterns than people who are not depressed.

We can test the above hypothesis by measuring the sleep patterns for people who are diagnosed as depressed and compare these patterns with those of people who are identified as not depressed. Sleep patterns could be operationalized in relation to sleep phases or simply as number of hours of sleep.

In stating the hypothesis, we were proposing a relationship between two variables—sleep patterns and depression—but the hypothesis says little about the nature of the relationship between them, only that they are related. If we ultimately find statistical support for the hypothesis that they are related, that relationship may mean that depression may influence sleep patterns, that sleep patterns may influence depression, or neither one. It could mean, simply, that the two variables naturally covary. But what if we had stated our hypothesis in a slightly different way?

> **Hypothesis 2:** Disturbed sleep patterns will cause depression.

Hypothesis 2 proposes a direct relationship between sleep patterns and depression—disturbed sleep patterns *cause* depression. In other words, one variable (sleep patterns) is predicted to affect the other (depression level). The variable that is predicted to do the affecting is known as the *independent* variable. In experimental research designs, this is the variable that we would either manipulate or introduce. For example, if we wanted to see whether disturbed sleep patterns cause depression, we could, in theory, disturb the sleep patterns of a number of people to see if they became depressed. (In reality, there are ethical reasons why we would do nothing of the sort!) Theoretically, however, sleep patterns would be the independent variable—the one that is manipulated.

The other variable in Hypothesis 2, depression, is known as the *dependent* variable. This is the variable that we believe is dependent upon the independent variable. According to the hypothesis, for example, a person's depression level (dependent variable) will depend on his or her sleep patterns (independent variable). However, we could turn the hypothesis around to read:

> **Hypothesis 3:** Depression will cause disturbed sleep patterns.

Here, depression, which does the causing, would be the independent variable. What is caused—sleep patterns—is the dependent variable. In order for a variable to be considered the dependent variable, it only needs to be believed to be affected in some way by the independent variable. Of course, other variables (referred to as intervening or extraneous variables, or some other related term) may help to cause the variation in the dependent variable that exists or to affect the relationship between the independent and dependent variables in other ways.

Note that the same variable, depression, can be considered independent or dependent according to the way the hypothesis is stated. In Hypothesis 2, depression was the dependent variable. In Hypothesis 3, it was the independent variable. Similarly, sleep patterns was the independent variable in Hypothesis 2, whereas it was the dependent variable in Hypothesis 3.

The terms *independent* variable and *dependent* variable are labels for the convenience of researchers and those with whom they communicate. They are used to indicate the direction of influence between or among variables, based upon certain logical evidence. They generally are used when the researcher is suggesting in a hypothesis that one variable is believed clearly to affect the other (more than vice versa). This was not the case in Hypothesis 1. Often in social work research we are not seeking to demonstrate that one variable directly influences another variable; we wish only to find support for the belief that variables (that we have neither introduced nor manipulated in any way) covary—that they are associated or correlated with each other. Putting it another way, we want to know if certain values of one variable naturally tend to be found with certain values of another variable. Our goal is to be able to predict the value of one variable by knowing the value of the other.

If demonstration of covariance and, ultimately, prediction are our goals, we use two other terms to describe the relationship between variables. The variable that we hope to use for prediction is referred to as the *predictor* variable. The variable whose values we hope to predict based upon values of the predictor variable is referred to as the *criterion* variable. Sometimes in our professional literature the terms *independent* and *dependent* have been used when the terms *predictor* and *criterion* more accurately describe the relationship that the researcher is attempting to demonstrate. In this book, we will be using the labels *predictor variable* and *criterion variable* in discussing those statistical analyses (i.e., correlation and depression) that in no way can be used to imply causality; the focus here is on their covariance. In other situations, where a form of analysis is discussed that is most frequently used to advance knowledge toward an understanding of causality, we will use the labels *independent variable* and *dependent variable*.

In the final analysis, the decision to label one variable as independent (or predictor) and the other as dependent (or criterion) comes down to a simple matter of logic. It all depends on what we believe to be the relationship between or among variables. In a hypothesis that seeks to explain the relationship between job satisfaction and the kind of supervision received by social workers, job satisfaction logically would be labeled the dependent variable and the kind of supervision received the independent variable. Variations in students' grades in a research course logically would depend on their motivation, on the amount of time they have studied, and on their basic intellectual capacity (grades =

dependent variable). Whether or not a student obtains a fellowship from a university may depend on the student's academic record and on the availability of fellowships, as well as on competition from other students (obtaining a fellowship = dependent variable).

In each of the previous examples, different values of the dependent variable are all logically dependent upon the different values of the independent variables. It would be illogical, for example, to believe that the direction of the relationship could be the reverse or to imply that the variables are merely associated or correlated and that no implication of causation can be inferred. If they are believed to be only correlated or associated, it would be more accurate to use the terms *predictor variable* and *criterion variable,* respectively, rather than the terms *independent variable* or *dependent variable.*

Social work researchers may be most interested in explaining differences in variables such as job satisfaction, grades, or success in obtaining a fellowship. Therefore, these are generally dependent variables in their research studies. Of course, researchers doing other studies, perhaps in other disciplines, may explore the reasons for the variations in kinds of supervision, student motivation, or methods employed to seek a fellowship. If so, such variables would become the dependent variables in those studies, and the researchers would use other factors as independent variables.

Variables are labeled as independent and dependent (or predictor and criterion) at the time that hypotheses are constructed. The labeling process assists the researcher both in selecting the appropriate statistical analysis and in interpreting the results of the analysis. It also helps the reader of a research report to understand what a researcher was attempting to accomplish through the statistical analysis.

LEVELS OF MEASUREMENT

The conceptualization and operationalization processes provide a necessary and orderly method for selecting and measuring variables. Formulation of hypotheses and labeling of variables within them (as either independent, dependent, predictor, or criterion) provide further specification of the researcher's focus and purpose.

Valid measurement of variables makes it possible for statistics to summarize research findings accurately and to analyze the relationships that appear to exist between and among them. However, before a statistical analysis can occur, we must make another determination about the methods and measurements used. The methods and measuring instruments used to measure variables can affect just how precisely we are able to measure them. Some variables, by their nature, cannot be precisely measured. Others can be measured precisely, but—through choice or accident—they are sometimes measured in a way (or with an instrument) that provides less precise data than could have been produced. For example, the variable *highest educational level attained* can be defined precisely by finding out the number of years, months, days, and even hours or minutes of formal education that research participants have experienced. Alternately, measurement of the variable may be as simple as learning a person's highest grade completed or even whether or not he or she ever attended college.

A researcher always considers measurement precision when operationalizing variables by selecting measuring instruments that will yield the highest level of measurement precision that is possible, given the study's contextual and methodological constraints. Prior to data analyses, a judgment must be made in reference to how precisely each variable has been measured. Determination of the variable's level of measurement is crucial because it provides direction as to the type of statistical analyses that can be undertaken. There are four levels of measurement a variable can take: (a) nominal, (b) ordinal, (c) interval, and (d) ratio.

Nominal

The first level of measurement is *nominal* measurement. It is the least precise level of measurement. Its categories (values) are discrete, or distinct, from each other. Nominal measurement is a classification system that categorizes variables into subclasses. Different values reflect only a difference in kind—nothing more. Because no implication of a quantifiable difference can be made, no rank ordering of values is possible. Variables such as gender, race, ethnicity, referral source, diagnosis, occupation, sexual orientation, marital status, and political party affiliation are usually regarded as nominal variables.

An example of a question in a research questionnaire that would produce only nominal measurement is:

Do you believe the government should subsidize the cost of abortions? (Circle one number below.)

1. Yes
2. No
3. Undecided

The requirements of nominal measurement are minimal. A nominally measured variable must have two or more categories (values), and the categories must be distinct, mutually exclusive, and mutually exhaustive. That is, each case (e.g., a research participant) must appropriately fit into only one of the categories, and there must be an appropriate category for each case. For example, there are only two classes of the nominal variable *life status*—living or deceased. These two categories are clearly exhaustive and mutually exclusive, as every person can be classified into one of the categories (exhaustiveness) but only one (exclusiveness).

In nominal measurement, numerals (or other symbols, such as letters) sometimes are assigned for convenience. Suppose we have divided the variable *type of intervention* into three categories: individual therapy, group therapy, and family therapy. We could assign numbers to the various types of intervention as indicated below:

1. Individual therapy
2. Group therapy
3. Family therapy

The numbers that we have used (i.e., 1, 2, 3) are merely value labels and serve only to classify. It would be meaningless in this case to say that 1 is more or less treatment than 2 or 3, or to make any other statement that implies that the three categories have any quantitative connotation. The value labels have no more meaning than if we had assigned letters of the alphabet to the different categories of intervention (e.g., A, B, C).

Ordinal

The second level of measurement is *ordinal* measurement. Ordinal measurement implies that a variable not only takes on different values but also that the values have some distinct quantitative meaning. With ordinal measurement, it is possible to rank order the values that the variable assumes from high to low or from most to least. Examples of ordinal variables are social class, occupational prestige, educational degrees received, ratings of client change, agreement on problem definition, ratings of treatment effectiveness, ratings of clients' satisfaction with treatment, and rankings of problem severity. Below are some of the more common value labels for variables at the ordinal level of measurement:

1. Considerable
2. Some
3. Little
4. None

1. High
2. Moderate
3. Low

1. Very effective
2. Somewhat effective
3. Somewhat ineffective
4. Very ineffective

1. Very satisfied
2. Somewhat satisfied
3. Somewhat dissatisfied
4. Very dissatisfied

1. Very severe
2. Severe
3. Mild
4. Very mild

An example of a question on a research questionnaire that would produce ordinal level measurement is:

How would you rate your social work supervisor? (Circle one number below.)

1. Very good
2. Good
3. Fair
4. Poor
5. Very poor

Value labels used with ordinal measurement make it possible to identify not only differences between variable subclasses but also their relative positions. In contrast, nominal measurement can only classify a variable into value categories that reflect simple differences in kind.

It is important to note that ordinal value labels neither indicate absolute quantities nor assume equal intervals between them. For example, we might have a social position scale that ranks social class according to a set of categories ranging from Class I (upper) to Class IV (lower). Since the classes do not necessarily represent equal intervals, we cannot say that Class I is exactly two class intervals higher than Class III or that this interval is exactly the same distance as the one that separates Class IV from Class II. The different values of an ordinal variable do not indicate either their absolute quantities or the exact distances that separate one category from another.

Interval

The third level of measurement is *interval* measurement. Like ordinal measurement, interval measurement also classifies and rank orders properties of variables; in addition, it places them on an equally spaced continuum. Unlike ordinal measurement, interval measurement has a uniform unit of measurement, such as one year, one degree of temperature, and so on. Therefore, value labels indicate exactly how far apart one value is from another. With an interval level variable, we can say that a research participant (or object) has "more" or "less" of a given property than another participant. In addition, we can also specify exactly how many units more or less.

With equal distances between the units, a measurement of 1 for a variable will be the same distance from a 4 (4 - 1 = 3) as a 6 is from a 9 (9 - 6 = 3). On a measuring instrument that is designed to measure intelligence, generally assumed to be quantified at the interval level, the difference between IQ scores of 100 and 105 (105 - 100 = 5) should reflect the same difference in intelligence between IQ scores of 115 and 120 (120 - 115 = 5). Two individuals with achievement scores of 50 and 60 (60 - 50 = 10) should have the same difference in achievement as exists between two individuals with scores of 80 and 90 (90 - 80 = 10).

An example of a question on a research questionnaire that would produce interval level measurement is:

What was your verbal score on your most recent Scholastic Assessment Test (SAT)? _____

Interval measurement does not have an absolute zero. This means that we cannot identify a point at which no quantity of the variable exists. We cannot say that a 2 is twice as much as a 1—only that it is one standard unit more. Since a reading of zero degrees on a Fahrenheit thermometer does not represent the absence of heat, a temperature of 60 degrees does not mean that it is twice as hot as a temperature of 30 degrees. Zero degrees Celsius is nothing more than a point (the temperature at which water freezes) arbitrarily chosen to receive the value label *zero*. The Fahrenheit and Celsius thermometers can generate only interval data (unlike a Kelvin thermometer, which has an absolute zero point).

Interval measurement indicates how far apart the values of a variable are from one another. It does not indicate the absolute magnitude of a property possessed by any particular person or object. This is possible only with the most precise type of measurement—ratio measurement.

Ratio

The existence of a fixed, absolute, and nonarbitrary zero point constitutes the only difference between interval and ratio measurement. Therefore, numbers on a ratio scale indicate the actual amounts of the property being measured. With such a scale, we can say not only that one person (or object) has so many units more of a variable than a second person, but that the first person has so many times more or less of the variable. Examples of ratio measurement are birth, death, and divorce rates; number of children in a family; and number of times that a client attended group treatment over a six-month period.

The absolute zero point has empirical meaning. All arithmetic operations are possible—addition, subtraction, multiplication, and division. This permits the valid use and meaningful interpretation of ratios formed by two or more scores. For example, a country with a birth rate of four children per couple has twice as high a birth rate as a country with a birth rate of two children per couple.

An example of a question on a research questionnaire that would produce ratio level measurement is:

How many times did you see your social worker during the past month?
_____

Most of the data used in social work practice and research are not at the ratio level. One way to test for the existence of ratio measurement is to think about the possibility of negative values for the variable being measured. If negative values can logically be assigned (e.g., a temperature of –25°F), then the measurement of the variable cannot be considered to be more than interval. With a ratio measurement, zero is assigned to the point at which no measurable quantity of the variable exists.

More on Levels of Measurement

As suggested earlier, and as will be discussed in more detail in Chapter 6, the level of measurement of variables is a major consideration in determining what form of statistical analysis to use. Whenever possible, we use a statistical method that uses all the measurement precision that we have available. However, sometimes statistical techniques appropriate for, say, interval data also require other assumptions about those variables and the way in which their values are distributed. If these assumptions cannot be met, statistical tests normally designed for use with ordinal or even nominal data should then be employed.

It is not correct to move in the other direction in the measurement hierarchy, that is, from the less precise to the more precise. If a variable is measured in a way that can be regarded as only nominal, it cannot be treated as ordinal since it lacks a natural ordering of its categories. Thus, it is not correct to use statistical tests designed for ordinal measurement with variables considered to be only nominal. Similarly, it is not appropriate to use statistical tests intended for use with interval or ratio measurement with variables considered to be only nominal or ordinal. This is important to remember because, as social workers increasingly are using standard statistical software computer packages for analysis of data, errors can easily be made. Currently, no computer software program is programmed to "catch" the fact that, for example, a researcher requesting a computer analysis designed for interval level data is using a variable that is only at the ordinal level. Incorrect analysis performed in this way can produce erroneous conclusions.

Levels of measurement usually refer to a judgment made by the researcher regarding the way in which a social phenomenon was conceptualized and operationalized. Less frequently the levels refer to inherent characteristics of the variables themselves. Depending on how a variable is conceptualized and operationalized in a given research study, a variable such as place of residence may be used to indicate the geographic place name of one's residence (nominal), the distance of that residence from a specific point on the globe (ratio), or the size of one's community (ordinal or interval).

In examining the nature of measurement of variables, certain other classifications (in addition to the level of measurement) are often useful. Like level of measurement, these classifications also guide the selection of the most appropriate methods for statistical analyses of data. Variables are classified as *discrete* if they can take on only a finite number of values (measurements), such as the number of correct answers on the Scholastic Assessment Test (SAT) or number of siblings. The opposite type of variable, which is called *continuous,* can theoretically take on any numerical value. The height of social work students or grade point average would be examples of continuous variables. If we were to take any two measurements of either variable, it would be possible theoretically that there could be one or more other measurements between them. The number of different values for the variables is unlimited, assuming that we can use instruments capable of measuring the values with ever-increasing precision.

A *dichotomous* variable is a specific type of discrete variable that can assume only one of two values. Examples are gender (male/female) or the result of an election (win/lose). Of course, we could take a more precisely measured variable—such as one that is interval or ratio level—divide its range of values into two groups (the top half and the bottom half), and convert it into a dichotomous level variable, such as older voters and younger voters. However, such an activity would not generally be beneficial for statistical purposes and, in fact, would be wasting the precision of measurement that is available for the variable *age.*

A special type of dichotomous variable is a *binary* variable. With binary variables, we customarily assign the numerical values 0 or 1 to indicate the presence (1) or absence (0) of something. For example, for the variable *car ownership* we would assign a value of 1 for people who own a car and a 0 for those who do not.

Another special type of dichotomous variable is known as a *dummy* variable. Suppose we wanted to take a nominal level variable like gender and make it more quantitative in order to allow us to perform a different type of statistical analysis. We could take the variable and convert it into two binary variables: female (1) and not female (0); or male (1) and not male (0). The two new variables thus created would be dummy variables, variables created from the data already collected in another format. In Chapter 12 we will see why creation of dummy variables can be a useful exercise.

TYPES OF STATISTICAL ANALYSES

There are several different ways in which the many different types of statistical analyses can be grouped. One way is according to their primary function or use. Statistical analyses involve methods for (a) designing and carrying out research studies, (b) summarizing and describing the major characteristics of collected data, and (c) making predictions or inferences about the likelihood that relationships between variables within the data set also exist beyond the data actually collected. This book will deal primarily with the last two uses of statistics, generally referred to as *descriptive* statistics and *inferential* statistics, respectively.

This is not to imply that the benefits derived from the use of statistics for designing and carrying out research studies are unimportant. Statistics are used, for example, to assist in the design, evaluation, and revision of data collection instruments and to determine what size research sample is likely to be representative of a larger number of cases (the population) from which it was drawn. In most of the discussion that follows, we will assume that these tasks have been performed well and that all that is needed is the statistical analysis of the data collected.

A second way of grouping statistical analyses relates to the number of variables that are involved. We can refer to statistical analyses as *univariate* (examining the distribution of values of a single variable), *bivariate* (examining the relationship between two variables), or *multivariate* (examining the relationship among three or more variables). Additional ways of grouping statistical analyses will be presented within the context of later discussions.

Descriptive Statistical Analyses

As we suggested earlier, *descriptive statistical analyses* are used to summarize the characteristics of data. These data may have been collected from a population (sometimes called the universe) of research participants (or objects). For example, such a population might be "all full-time students currently enrolled in accredited schools of social work." The statistical characteristics of these students are called *parameters.* Descriptive statistical analyses also are used to summarize the characteristics of a research sample. When we are talking about a research sample, its characteristics are called *statistics.*

After data on the members of a particular population or sample are collected, the original measurements, or scores (called raw data), frequently can be overwhelming. A way must be found to organize and summarize the most important, salient characteristics of the data. Through the use of descriptive statistical analyses (also known as data reduction), we can derive summaries of information. Descriptive statistical analyses are often preliminary to other (inferential) statistical analyses, but in some types of research—surveys and some qualitative research designs, for example—they are the primary focus of the data analysis.

Descriptive statistical analyses are based on measurements actually taken of the sample or population. In using them, our concern does not extend beyond the particular research participants studied. It generally consists of graphs, tables, and descriptive numbers, such as averages and percentages—all of which are easier to comprehend and interpret than a long list of data reporting the results of measurement of each variable for every case. The main purpose of descriptive statistical analyses are to reduce the whole collection of data to simple and more understandable terms without distorting or losing too much of the valuable information collected. Of course, any summary sacrifices some detail, and descriptive statistical analyses are no exception.

Inferential Statistical Analyses

Inferential statistical analyses are used when we have access to only a sample drawn from a population and when we do not have in our possession all the raw scores that could theoretically exist in the total population. It consists of procedures for determining how safe it would be to make generalizations about characteristics of population parameters based on measurements drawn from a sample (statistics). The sample statistics that we have are regarded as merely estimates of what all measurements of the population might look like (the population parameters). The question is, how accurate are these estimates?

Inferential statistical analyses frequently are used when we seek to demonstrate support for hypotheses. We will look at this specialized use in greater detail in Chapter 5.

CONCLUDING THOUGHTS

In this chapter we have stressed that sound measurement is a prerequisite to the meaningful statistical analyses of data. We have reviewed basic concepts vital to an understanding of the measurement of variables. The reader can find them discussed more thoroughly in texts used in graduate and undergraduate social work research methods courses.

As we suggested in this chapter and will demonstrate in the chapters that follow, statistical analyses involve methods for gathering, organizing and summarizing, analyzing, and evaluating data. It is not, or should not be, some mysterious mathematical process. In fact, statistical analyses are little more than the application of logic and common sense reasoning to the analysis of data.

STUDY QUESTIONS

1. Discuss why good measurement is essential to meaningful statistical analyses. Use an original example in your discussion.
2. Discuss how a variable differs from a constant. Provide an original example of each in your discussion.
3. In a research hypothesis, what do we call the variable whose variations we are most interested in explaining? What do we call the variable that we believe may affect these variations? Which other terms are substituted if we are primarily interested in predicting the value of one variable through knowing the value of the other? Provide original examples to illustrate your understanding of these terms.
4. What additional characteristic does a valid measurement possess that one that is merely reliable may not?
5. What are three different ways that we might operationalize the variable *motivation to attend a graduate school of social work*?
6. What does the term *value* mean when referring to a variable? Provide an original example in your discussion.
7. What additional criterion must be met for a variable to be considered ordinal that is not a requirement for nominal measurement? Provide an original example in your discussion.
8. What is required for ratio measurement that is not required for interval measurement? Provide an original example in your discussion.
9. Operationalize the variable *educational level* so that it would produce nominal level measurement, ordinal level measurement, interval level measurement, and ratio level measurement.

10. In your own words, discuss the differences between descriptive and inferential statistical analyses. Describe some social work situations where the use of each would be appropriate. What does inferential statistical analyses attempt to determine that descriptive statistical analyses do not? What other methods are used to classify different types of statistical analyses?

11. Describe ways that we can use statistical analyses in social work practice, in social work education, and in social work research.

12. Discuss the difference between values of a variable and frequencies of the values of a variable. Provide an original example in your discussion.

13. Discuss the role that conceptualization and operationalization have in social work practice and research.

14. Construct a hypothesis with one nominal independent variable and one nominal dependent variable. Explain how you would measure the dependent variable to produce the desired level of measurement.

15. Construct a hypothesis that has one nominal independent variable and one ordinal dependent variable. Explain how you would meaasure the dependent variable to produce the desired level of measurement.

16. Construct a hypothesis that has one nominal independent variable and one interval dependent variable. Explain how you would measure the dependent variable to produce the desired level of measurement.

17. Construct a hypothesis that has one nominal independent variable and one ratio dependent variable. Explain how you would measure the dependent variable to produce the desired level of measurement.

18. What level of measurement is the variable *highest social work degree received*? Justify your response. What other ways of operationalizing the variable would produce different levels of measurement? Explain.

19. Find a research-based article in a social work professional journal that is of some interest to you. Answer the following questions in relation to the article:

    a. How much of the article is a report of statistical analyses per se (as opposed to theory, ideas, implications, research design, sampling, data collection, and so on)?

    b. Was the study conducted using a population or a sample? If a sample was used, how was it selected? Do you believe the findings can be generalized to the population from which it was drawn? Why or why not? Discuss.

    c. Do you feel the author conceptualized the key variables correctly? Why or why not? How could the author have conceptualized them differently? Provide examples.

    d. Do you feel the author operationalized the dependent (or criterion) variable correctly? Why or why not? How could the author have operationalized it differently? Provide examples.

    e. What were the study's independent (or predictor) and dependent (or criterion) variables? What level of measurement were they? Justify your response.

    f. What statistical method(s) was used in the article? Was it descriptive and/or inferential analysis?

    g. Do you think the measurements of the key variables were reliable and valid? Why or why not?

    h. Did the author use a standardized measuring instrument to measure the dependent (or criterion) variable? If so, which one was used? Do you feel the instrument measured what it was supposed to measure? Why or why not?

chapter 2

Frequency Distributions and Graphs

The data collected in research studies need to be organized and summarized. There are two primary formats in which this can be done: (a) through the use of tables in the form of frequency distributions, and (b) by summarizing the data in graphical form. This chapter shows how these two formats can be helpful in visualizing the distribution of the values of variables within a research sample or population.

FREQUENCY DISTRIBUTIONS

One of the first questions often asked after data have been collected relates to how many persons (or objects) fell into each value category for every variable that was measured. We are curious to know how the research sample or population "broke"— that is, what the frequency was for each value of each variable. All frequency distribution tables are designed to provide an easy answer to this question.

If a variable is at the nominal level, frequency distributions are constructed directly from raw data, but if data are at the ordinal level of measurement or higher, it is often helpful to first arrange them into an array. An *array* is an ordering of every case value within raw data from the lowest (smallest) value that occurred to the highest (largest). A hypothetical research example will be used to show what an array looks like and to illustrate various frequency distributions that can be formed from it.

Suppose that a social work agency administrator wonders whether the agency is truly serving "older" residents of the community as written in its mission statement. (The agency has operationally defined "older residents" as 50 years of age or older.) The administrator decides to record the ages of all new clients who

TABLE 2.1   Raw data: Clients' names and ages

Name	Age	Name	Age
Rashad	32	Rosemarie	37
Rosina	27	Marguerite	49
Brad	26	Raquel	31
Chuck	21	Peter	27
Shanti	37	Clarisse	37
Kathy	31	Karen	26
Antoinette	32	Elwin	49
David	69	Tony	21
Herb	26	Leon	27
Vincent	31	Mario	31

apply for services every Tuesday in October (the research sample). Twenty clients apply for services during this month, and their ages are obtained from the agency's intake forms. These raw data for the 20 clients are then listed, as in Table 2.1.

As we can see from the data in Table 2.1, the first new client was Rashad, who was 32 years of age; the second client was Rosina, 27 years of age; and so on. The raw data from Table 2.1 can be placed in an array, such as Table 2.2.

Note that the array displays the agency data from the lowest value (21) to the highest value (69). Every one of the clients is represented by a number—his or her age. Table 2.2 demonstrates that 2 of the 20 clients were 21 (low) years of age (Chuck and Tony) and only 1 was 69 (high) years of age (David).

The data in Table 2.2 provide a beginning answer to the research question about clients served. Only one client (David) meets the agency's operational definition of "older," since he was the only client over 50 years of age. As we can see, Table 2.2 makes it much easier for us to "eyeball" the data compared to Table 2.1. If the data had consisted of 250 case values instead of just 20, the formation of an array would have been even more helpful in this regard.

Having formed an array with the data, it is now possible to construct frequency distribution tables in order to make them more meaningful. A frequency distribution will consolidate the data taken from an array such as Table 2.2.

Absolute Frequency Distributions

To construct an *absolute frequency distribution* (also known as a simple frequency distribution), we simply count the number of times each value for the variable was found to occur and place it in a table next to that value. An absolute frequency distribution may be constructed for data at any level of measurement.

Table 2.3 reports that the clients' ages in our example ranged from 21 (Chuck and Tony) to 69 (David) and that the age most frequently reported was 31 (Kathy, Vincent, Raquel, and Mario). The absolute frequency column on the right side

TABLE 2.2 Array: Clients' names and ages (from Table 2.1)

Name	Age	Name	Age
Chuck	21	Raquel	31
Tony	21	Mario	31
Brad	26	Rashad	32
Herb	26	Antoinette	32
Karen	26	Shanti	37
Rosina	27	Rosemarie	37
Peter	27	Clarisse	37
Leon	27	Marguerite	49
Kathy	31	Elwin	49
Vincent	31	David	69

of the table indicates the number of times each value occurred. For instance, Chuck and Tony were 21 years of age, and as a group they constitute a frequency of 2—that is, the absolute frequency for the value 21 is 2. Similar data are given for each of the eight ages that occurred. Absolute frequency distributions sometimes are seen in research reports, but, more commonly, they appear as just part of the more complex descriptive distributions that will be discussed next.

Cumulative Frequency Distributions

It is possible to provide additional description of data by adding another column to an absolute frequency distribution. A *cumulative frequency distribution* table such as Table 2.4 can be constructed if the data are at the ordinal level of measurement or higher (that is, if an array can be formed as in Table 2.2).

TABLE 2.3 Absolute frequency distribution table: Clients' names and ages (from Table 2.2)

Name	Age	Absolute Frequency
Chuck + Tony	21	2
Brad + Herb + Karen	26	3
Rosina + Peter + Leon	27	3
Kathy + Vincent + Raquel + Mario	31	4
Rashad + Antoinette	32	2
Shanti + Rosemarie + Clarisse	37	3
Marguerite + Elwin	49	2
David	69	1
Total		20

TABLE 2.4   Cumulative frequency distribution table: Clients' names and ages (from Table 2.3)

Name	Age	Absolute Frequency	Cumulative Frequency
Chuck + Tony	21	2	2
Brad + Herb + Karen	26	3	5
Rosina + Peter + Leon	27	3	8
Kathy + Vincent + Raquel + Mario	31	4	12
Rashad + Antoinette	32	2	14
Shanti + Rosemarie + Clarisse	37	3	17
Marguerite + Elwin	49	2	19
David	69	1	20

As Table 2.4 shows, 2 clients were 21 years of age and 3 clients were 26. Thus, the cumulative frequency of the clients' ages 26 and under is 5 (2 + 3 = 5). We can also see that 17 clients (2 + 3 + 3 + 4 + 2 + 3 = 17) were 37 years of age and under. In a cumulative frequency distribution, the last number in the cumulative frequency column always is the same as the total number of cases, indicating that all case values have been included.

Percentage Distributions

A third type of frequency distribution table, the *percentage distribution* table, includes other information about our data. Table 2.5 uses the same data as Table 2.4, but it adds a percentage column alongside the cumulative frequency column.

Since there are 20 clients in the sample, each client represents 5 percent of the sample (100/20 = 5%). The number in the percentage column for each age

TABLE 2.5   Percentage distribution table: Clients' names and ages (from Table 2.4)

Name	Age	Absolute Frequency	Cumulative Frequency	Absolute Percent
Chuck + Tony	21	2	2	10
Brad + Herb + Karen	26	3	5	15
Rosina + Peter + Leon	27	3	8	15
Kathy + Vincent + Raquel + Mario	31	4	12	20
Rashad + Antoinette	32	2	14	10
Shanti + Rosemarie + Clarisse	37	3	17	15
Marguerite + Elwin	49	2	19	10
David	69	1	20	5

TABLE 2.6 Cumulative percentage distribution table: Clients' names and ages (from Table 2.5)

Name	Age	Absolute Frequency	Cumulative Frequency	Absolute Percent	Cumulative Percent
Chuck + Tony	21	2	2	10	10
Brad + Herb + Karen	26	3	5	15	25
Rosina + Peter + Leon	27	3	8	15	40
Kathy + Vincent + Raquel + Mario	31	4	12	20	60
Rashad + Antoinette	32	2	14	10	70
Shanti + Rosemarie + Clarisse	37	3	17	15	85
Marguerite + Elwin	49	2	19	10	95
David	69	1	20	5	100

that occurred within the sample of clients represents the absolute percentage of the entire sample (20 clients) who were that age. As Table 2.5 indicates, 2 people (Chuck and Tony) were 21 years of age, and together they represent 10 percent of the total number of clients (5 percent for Chuck + 5 percent for Tony). Similarly, Brad, Herb, and Karen together represent 15 percent of the total sample (5 percent for Brad + 5 percent for Herb + 5 percent for Karen). Of course, the total for all the clients always equals 100 percent.

Cumulative Percentage Distributions

A fourth type of frequency distribution table is the *cumulative percentage distribution* table. It combines features of both Table 2.4 and Table 2.5, but it contains an additional column that reports cumulative percentages for each of the case values for a variable that occurred in the sample. In our example, it would tell us what percentage of all 20 cases were a given age or below.

As shown in Table 2.6, for instance, 2 clients (Rashad and Antoinette) were 32 years of age, and together they represent 10 percent of all clients (5 percent for Rashad + 5 percent for Antoinette). Additionally, 70 percent (14/20 = 70%) of all clients were 32 years of age or younger.

GROUPED FREQUENCY DISTRIBUTIONS

Sometimes it is difficult to interpret frequency distribution tables because of the unequal range between the values of the variables that occur within a sample or population. In our example, the variable *age* is distributed in such a way that there are different size "gaps" (e.g., 21 to 26, 27 to 31, 32 to 37, and 49 to 69). It is sometimes easier to visualize and to comprehend the meaning of these data if they were "condensed" into a smaller number of standardized value categories or groupings (e.g., 20 to 29, 30 to 39, and so on). These groupings could then be displayed using any of the frequency distributions in Tables 2.3 to 2.6, or we might choose to use only those columns from Table 2.6 that are of special interest to us. In creating grouped frequency distributions we can use the age groupings in the first column instead of the clients' actual ages and adjust the numbers in the other columns accordingly. Table 2.7 is a grouped cumulative percentage distribution of the data that we have been using. Note that it does not contain the actual frequency for any one age, but that it nevertheless provides a good overview of how the data were distributed in the sample.

Grouped frequency distributions are especially useful when there are too many different values for a variable to list each of them in the form of a nongrouped frequency distribution. This often occurs when there are a large number of cases and when forming frequency distributions for variables that are at the interval or ratio levels of measurement. For example, the number of miles driven to class by students in a school of social work would make for a lengthy list, especially if miles were measured in fractions or tenths of a mile. Transforming the observations into meaningful groupings may make it easier for the reader to visualize the distribution of the data.

What is a "meaningful" grouping when we refer to grouped data? It is a grouping that reduces the number of categories of a variable to a reasonable number while not losing any more measurement precision than is necessary. Whenever possible, groupings should encompass an equal or nearly equal number of potential case values. The groupings also should reflect homogeneity for the variable, which means the groupings should use value intervals small enough that cases within them look similar in relation to the variable. In our example, meaningful value groupings might be:

    5 miles or less
    6-10 miles
    11-15 miles
    16-20 miles
    21 miles or more

Note that all persons in the "5 miles or less" category share a similar characteristic—they all live relatively close to the university. Those in the 6-10-mile grouping live fairly close, but they probably would not want to walk to class, and so on. There is a logic to the groupings—persons within them all share some similarity with the other members.

Of course, every case should fall cleanly into one and only one grouping. In the previously mentioned groupings, a consistent method would have to be devised for assigning those students who would be on the edge of the different

TABLE 2.7 Grouped cumulative percentage distribution table: Clients' ages (from Table 2.6)

Ages	Absolute Percentage	Cumulative Percentage
20–29	40	40
30–39	45	85
40–49	10	95
50–59	0	95
60–69	5	100

value groupings (e.g., 10.6 miles) into a group. A customary way to do this, for example, is to consider 11 miles as really representing the interval 10.50 to 11.49 miles. Then the student who drives 10.6 miles would fall into only the 11–15-mile grouping.

USING FREQUENCY DISTRIBUTIONS TO ANALYZE DATA

Frequency distributions sometimes can be very revealing. Cumulative frequency distributions are especially useful when we are interested in knowing approximately where a particular value fell relative to the other values in a distribution of values. Suppose, for example, that the administrators of a large social service organization want to study the problem of unauthorized staff absenteeism. They would like to identify seasonal patterns that may exist that could possibly be reduced by creating new policies on vacations and annual leave. A cumulative frequency distribution table, such as Table 2.4, or a cumulative percentage distribution table, such as Table 2.6, might be used.

Tables 2.8 and 2.9 indicate that in April, absenteeism occurred 30 times (Table 2.8), or only 15 percent (Table 2.9) of the total amount of absenteeism

TABLE 2.8 Cumulative frequency distribution table: Staff days lost by month at XYZ agency

Month	Monthly Totals	
	Absolute Frequency	Cumulative Frequency
April	30	30
May	40	70
June	60	130
July	70	200

TABLE 2.9 Cumulative frequency
distribution table: Staff days lost by month
at XYZ agency

| | Monthly Totals | |
| | Absolute | Cumulative |
Month	Frequency	Frequency
April	15	15
May	20	35
June	30	65
July	35 .	100

for the 4-month period. The cumulative amount of absenteeism was 200 days
(Table 2.8). Only 35 percent (Table 2.9) occurred during April and May, and the
other 65 percent (100% – 35% = 65%) occurred during the months of June and
July. The two tables would seem to suggest to the agency administrators that there
is a seasonal pattern of absenteeism.

If we want to compare measurements taken from two different groups or
data sets, frequency distributions also can be helpful. If the measurements taken
on the groups differ somewhat (e.g., if the actual ages were recorded for one
data set and age ranges were recorded for the other one), grouped frequency
distributions can be used to make the different data sets comparable.

Example

An example will illustrate how frequency distributions can be used to compare
two groups of data sets. Sue, a social worker, developed a state merit examination study guide. She wished to get a preliminary indication of whether or not
it was effective. (Later we will discuss how more sophisticated forms of statistical analyses could provide more definitive answers about the study guide's effectiveness.) She decided to look at the respective scores of persons who used the
study guide (experimental group) and persons who did not use it (control group).
Cumulative distribution tables displaying the differences between the results for
the two groups are shown in Tables 2.10 and 2.11.

TABLE 2.10 Grouped cumulative
percentage distribution table: Experimental
group's scores ($N = 300$)

| | Absolute | Cumulative |
Scores	Percentage	Percentage
50–59	0	0
60–69	10	10
70–79	40	50
80–89	30	80
90–100	20	100

TABLE 2.11 Grouped cumulative percentage distribution table: Control group's scores ($N = 200$)

Scores	Absolute Percentage	Cumulative Percentage
50–59	5	5
60–69	15	20
70–79	40	60
80–89	35	95
90–100	5	100

As we can see from Table 2.10, 40 percent of the people in the experimental group scored between 70 and 79 on the examination, and 50 percent scored 79 or lower. In Table 2.11 we note that 40 percent of the people in the control group scored between 70 and 79 on the examination, and 60 percent scored 79 or lower.

We can note that cumulative percentages also make it possible to calculate at least approximate percentile ranks for individuals within the two subsamples. *Percentile ranks* indicate the percentage of the cases within a group whose values fall below a particular value. Suppose that a particular individual in the experimental group scored a 90 on the merit examination after using the study guide. A review of Table 2.10 would indicate that person scored higher than at least 80 percent of all persons in the experimental group, or that the test taker scored at approximately the 80th percentile. Percentile ranks enable us to put an individual score in perspective relative to the other scores in a group. In Chapter 4, we will present a more precise method to figure the percentile for a given case value within a group of values.

Also notice that the two groups contained different numbers of cases (N), 300 in the experimental group and 200 in the control group. Using percentages facilitates drawing comparisons between two or more groups of unequal size. This can be very helpful if two research samples are either of unequal size from the beginning or, if they begin equal, become unequal because more people are lost from one group than from the other.

MISREPRESENTATION OF DATA
USING FREQUENCY DISTRIBUTIONS

From a statistical perspective, the two subsamples of 200 and 300 research participants in the previous illustration were fairly comparable in size. Percentage comparisons drawn from them would be appropriate, since they would make the data easier to interpret. But a word of caution is in order: The practice of drawing comparisons between two groups of vastly unequal sizes actually can distort rather than clarify the data for the reader. Another example will be used to demonstrate how this can happen.

Example

Emma, a social agency administrator, proudly reported to the board of directors that the results of the agency's affirmative action plan for the years 1994–1995 were "outstanding." She took the data from Table 2.12 and presented them to the board of directors as a glowing endorsement of the agency's efforts to hire women. She noted that "in five of the six job classifications (i.e., A, B, C, E, and F), we have hired a higher percentage of women than men." Emma was able to make this statement because she was using percentages with subcategories of very different sizes. The actual data she summarized present a drastically different picture. In fact, as Table 2.12 illustrates, Emma's agency hired 78 percent of all male applicants but only 21 percent of all female applicants.

We all know that it is possible to "lie with statistics," but more often than not the misrepresentation is not deliberate. Emma's report may have been an honest mistake based on an inadequate understanding of the importance of using comparably sized subsamples when making comparisons using frequency distributions. There is a way to avoid giving the impression that one is attempting to distort the facts in such situations. The actual numbers on which percentages are based can be reported along with the percentages for all subgroups.

Percentages are helpful to others in comprehending large numbers, especially when they are used to report odd numbers, such as 146 out of 411 (35.5 percent). However, they are generally meaningless, if not misleading, in reporting data from small samples. There is little value in reporting, for example, that "60 percent of the graduates of an intensive job training program found work" if the 60 percent really means three of five completed the program (3/5 = 60%). With a small number of cases, it is best not to report percentages. For one thing, small numbers are quite comprehensible by themselves, and reporting them as percentages can be misleading, since percentages may subconsciously make us think in terms of larger numbers, like 100 or more.

TABLE 2.12 1994–1995 Hiring data for XYZ agency broken down by gender

Classification	Males Number	Percent	Females Number	Percent
A	3 of 6	50	4 of 6	67
B	1 of 3	33	1 of 2	50
C	0 of 1	0	1 of 10	10
D	85 of 100	85	2 of 40	5
E	2 of 3	67	2 of 2	100
F	3 of 7	43	4 of 7	57
Totals	94 of 120	78	14 of 67	21

GRAPHICAL PRESENTATION OF DATA

Sometimes it is be difficult to grasp the overall meaning of frequency distribution tables, but a picture can communicate the overall meaning of data almost immediately. However, when graphic presentations of data are used effectively, data are displayed in such a way that "the bigger picture" becomes readily apparent.

Graphic representations generally sacrifice detail in an effort to improve communication, but the sacrifice is justifiable and even desirable in many situations. Graphs are useful for displaying the findings of a research study that involves a large number of cases. If the intended audience is not research oriented, they may become impatient or uninterested with tabular presentation of vast amounts of data; graphs are more likely to hold their interest. Also, if it is essential to get a point across quickly and dramatically, graphs can do the job. They allow an audience to get a comprehensive picture of the distribution of the values of a variable without having to focus on unnecessary detail, such as the values of individual cases.

Like all methods of displaying data, pictorial representations can be constructed so as to produce misleading statements. It is possible to "lie with graphs," just as with other types of statistical analyses. This fact should serve to alert us to the dangers inherent in the use of graphs and cause us to ask whether they are communicating the findings of research studies accurately, but it should not preclude their use.

There are several graphs that commonly are used to display how many cases (persons or objects) were found to have the various measurements of a variable. Both the measurement level of the variable and clarity of data portrayal determine which of the various options is best. Most graphs are drawn (usually with computer software) using the traditional x- and y-axes normally introduced in elementary school mathematics courses. The vertical line is known as the y-axis, and the horizontal line is known as the x-axis. The point where the x-axis and y-axis meet (see Figure 2.1) is called the *point of origin.* In a graph used to present a description of the values observed for one variable, the y-axis is used to indicate frequencies for each value. The graph may extend along the x-axis to the left of the y-axis if there are negative values of the variable.

Bar Graphs

A basic method for organizing nominal data and representing them in pictorial form is the *bar graph,* also called a bar chart. Bars of equal width are drawn so that they do not touch. This suggests the qualitative (not quantitative) differences in values of nominal data. Figure 2.2a is an example of a simple bar graph.

If lines rather than bars are used, they are drawn so their length reflects the frequencies with which given values occur. We refer to this type of graph as a *line diagram.* Line diagrams may be constructed simply using vertical lines instead of the bars of a bar graph, or they may be constructed so the lines run

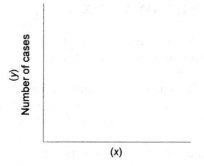

FIGURE 2.1 Basic outline of a graph for depicting values for two variables

parallel to the *x*-axis (i.e., horizontally) with different nominal values placed along the *y*-axis, as in Figure 2.2b. The choice of whether to use a bar graph or a line diagram is really one of esthetics.

Pie Charts

At times, when the various values of a nominal variable add up to a whole, we may choose to use a *pie chart,* also called a pie graph. The components of a pie chart reflect segments of the whole. Pie charts are usually circle graphs divided into wedges representing fractions of the total circle (the pie). For example, if we were showing how a client's family budget was divided into sums for food, shelter, clothing, and recreation, the total budget could be displayed as a circle or, given the availability of computer graphics, even as a bag of money. Pieces of the pie (or the bag) would be sized to reflect the portion of the total budget represented by the various budget items.

Figure 2.3 is an example of a pie chart that portrays the percentages of staff who work at XYZ agency broken down by their highest postsecondary social work degree obtained. Note, for example, that those staff members whose highest degree completed is an associate degree (25 percent of all staff) occupy one-fourth the total area of the "pie." The angle at the center of the pie for its associate degree segment is 90° (360° × .25). The angle for staff members who hold a doctorate is 36° (360° × .10).

Pie charts provide a way to make a rapid visual appraisal of the distribution of values. Their main limitation is that they cannot easily accommodate very many different values of a variable without becoming too complicated or even illegible. Figure 2.3 is quite comprehensible, but a pie chart that displays the frequencies for all of the academic majors within a major university would be impractical. There would have to be many more "slices of the pie," some of them so small as to be barely visible.

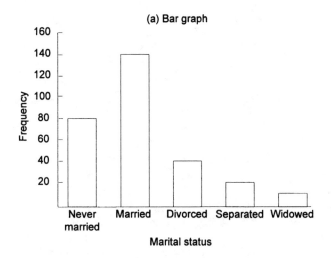

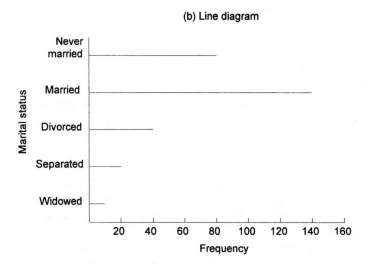

FIGURE 2.2 Two similar graphs: Bar graph and line diagram portraying the marital status of active clients in XYZ agency

Histograms

A useful graph for displaying ordinal data is called a *histogram*. Histograms look like bar graphs except the bars touch. The rank order of the variable's categories determines the sequence of the values in the graph. The bars of a histogram displaying ordinal data are of equal width, such as those in Figure 2.4, which displays the data derived from a review of February case records at a social agency.

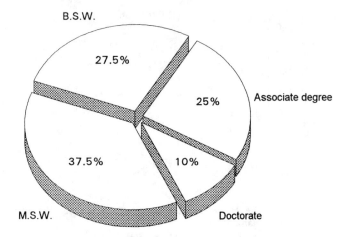

FIGURE 2.3 Distribution of social workers in XYZ agency broken down by highest social work degree attained

A histogram, like a bar graph, uses the height of a bar to display the frequency of a value for a given variable. A comparison of the frequency for different values is implicit. For example, if a bar of one length represents the frequency for one value, a bar twice as long represents a frequency twice that large for another value.

A histogram also may be used to display interval or ratio data. If grouped frequencies of unequal intervals are being displayed, the bars may be constructed so that their different widths correspond to the size of the different intervals. Figure 2.5 illustrates this variation on the usual form of a histogram. Note that the graph accurately portrays both frequencies by showing bars of different heights, intervals, and widths. To do this, it must employ the measurement precision available within interval or ratio data that is not present within ordinal data.

Frequency Polygons

After constructing a histogram using interval or ratio data, we can convert the histogram into a *frequency polygon*. Frequency polygons are designed to portray the overall shape of a distribution of scores (or values). If we were to take a pencil and mark a dot in the middle of the top of each vertical bar in a histogram and then connect the dots with a straight line, we would have a frequency polygon. Lines usually are drawn at each end of the distribution of values to connect it to the horizontal axis. Figure 2.6 is a frequency polygon displaying data that might have been collected at intake and stored in an agency computer-assisted management information system.

FIGURE 2.4 Histogram: Frequency of discharge status at XYZ agency during February

The data in Figure 2.6 probably were collected as interval or ratio level (e.g., as actual dollars and cents of income reported on clients' most recent federal income tax returns). When they were grouped into income intervals to reduce the number of values (categories) of the variable, some sacrifice of accuracy (measurement precision) occurred. The variable became only ordinal. As Figure 2.6

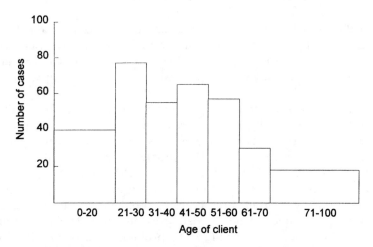

FIGURE 2.5 Histogram: Ages of clients on record in XYZ agency during October

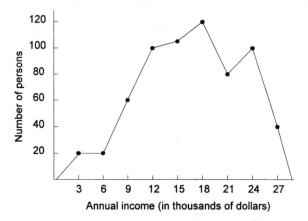

FIGURE 2.6 Frequency polygon: Annual income of families receiving family counseling at XYZ agency (rounded to nearest category)

demonstrates, it is possible to plot a frequency polygon for data that are only ordinal level, but this should be done only when the data are inherently interval or ratio. However, frequency polygons present a more accurate portrayal of the distribution of a variable when used with data that are at the interval or ratio level of measurement.

A COMMON MISTAKE IN DISPLAYING DATA

Even relatively simple ways of displaying data like bar graphs and histograms easily can be made quite complicated. Unfortunately, as we become more creative and attempt to display greater and greater amounts of data, results sometimes become more difficult to interpret. We will use an example to illustrate this important point.

The horizontal bars in bar graphs may be extended to the left or to the right simultaneously. Figure 2.7 displays data relative to the presenting problems of clients in a human service agency. It illustrates the actual percentage of clients who experience each of three presenting problems—environmental, psychological, and social—as identified by social workers at XYZ agency.

The graph portrays the realities of a situation in which a given client may have displayed one, two, or all three types of problems, but it takes quite a bit of study to be able to understand the data presented in it. The x-axis represents the percentage of clients displaying each problem area (right side) and the percentage not displaying the problem area (left side). Thus, the total number of clients either displaying or not displaying each problem must equal 100 percent. With some study, Figure 2.7 indicates that 30 percent of all clients had

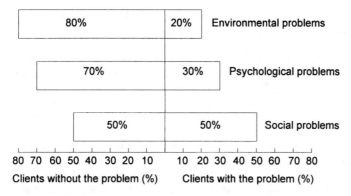

FIGURE 2.7 Bar graph: Types of client problems at XYZ agency

psychological problems and, of course, 70 percent did not; 50 percent had social problems, while the other 50 percent did not; and so forth.

CONCLUDING THOUGHTS

The data in Figure 2.7 could have been communicated more easily using three simple bar graphs, one for each type of presenting problem. With currently available computer assistance, it is possible to develop some very complex frequency distributions and to create some very intricate graphs, only a few of which have been mentioned in this chapter. However, this technology can be a mixed blessing. As we are tempted to use more complex and creative frequency distributions and graphs, we must always ask ourselves, will they really help the reader understand our data? Or will they have the undesirable effect of confusing the reader? After all, we must remember that *communication* is the goal of both frequency distributions and graphs. When we end up confusing the person who is trying to understand what the data looked like, we have failed to accomplish this goal.

STUDY QUESTIONS

1. Discuss how an array differs from raw data. Provide an original example in your discussion.
2. What additional information is conveyed in a cumulative frequency distribution that is not present in an absolute frequency distribution? Provide an original example in your discussion.
3. What type of frequency distribution would tell us what percentage of AFDC clients in a county social service agency have fewer than four children?

4. In a study attempting to relate type of counseling to employment, why would it be inadvisable to group the variable *number of interviews* as 1-10, 11-20, and over 20?

5. Why is it misleading to report a 50 percent success rate in a treatment program for alcoholics when there were only eight people in the treatment program?

6. How does a bar graph differ from a histogram? Provide an example of each.

7. If an agency with an $800,000 annual budget allocates $160,000 for travel expenses, what portion (percentage and number of degrees) of a pie chart would be reserved for the travel segment?

8. Why are frequency polygons an accurate portrayal of data only if data are at the interval or ratio level?

9. Why is simpler usually better when selecting a graph to display data?

10. Describe several ways that we could use graphs to display changes in the ethnic composition of an agency's professional staff between 1988 and 1995.

11. At the University of Twin Peaks, the number of males and females in various major fields of study are as follows:

Major	Number of Males	Number of Females
Social Work	20	80
Humanities	40	40
Business	60	50
Education	90	90
Nursing	10	90

   a. What percentage of social work majors are female?
   b. What percentage of the total student body are male?
   c. Construct a five-slice pie chart for the different female majors. Do the same for the males.
   d. Construct a bar chart like Figure 2.2a for all males broken down by major. Do the same for the females.
   e. What level of measurement is the variable *major*? Justify your response.

12. The dean of your school of social work has requested that you organize and present data from a two-question (two-variable) survey to see if student satisfaction in the social work program might relate to age of students. There are a total of 220 students in the program; 55 of them (25 percent) were randomly selected to receive the survey. One of the research questions was stated as: "How satisfied are you with the social work education you are currently receiving?" Possible responses were:

   1. Very satisfied
   2. Satisfied
   3. Somewhat satisfied
   4. Somewhat dissatisfied
   5. Very dissatisfied

   a. At what level of measurement is the variable *satisfaction*? Justify your response.
   b. Is the variable the independent variable or dependent variable? Justify your response.

c. The raw data for the variable were as follows: 1, 4, 5, 3, 2, 1, 4, 5, 5, 4, 2, 5, 1, 1, 2, 3, 4, 3, 3, 3, 3, 3, 4, 5, 4, 3, 3, 2, 2, 1, 3, 4, 2, 2, 2, 4, 2, 1, 1, 2, 3, 4, 5, 4, 3, 2, 1, 1, 2, 2, 3, 2, 2, 1, 1, and 3.

d. How many students completed the survey? Should these students be considered a research sample or a population? Why?

e. Construct an absolute frequency distribution to display these data.

f. Construct a cumulative frequency distribution to display these data.

g. Construct a percentage distribution to display these data.

h. Construct a cumulative percentage distribution to display these data.

i. Do you believe that, overall, the students were satisfied with the social work education they were receiving? Justify your response.

j. The dean took the results to the vice-president and said that the social work students were very pleased with their education. Was the dean correct in saying this given the results of the survey? Why or why not?

chapter 3

Central Tendency and Dispersion

In Chapter 2 we presented simple tabular and graphic presentations of data that display how values of a variable are distributed within a sample or population. There will be times, however, when we will want to go beyond simply displaying data in an effort to direct the reader's attention to some specific characteristic of these data. We may want to summarize the data by reporting on what was found to be a "typical" attribute of them and/or to what degree the values of a variable differed from the "typical value."

WHAT IS "TYPICAL" IN STATISTICAL TERMS?

In our everyday language, we tend to use the word *typical* rather loosely. We speak of the "typical" client or the "typical" starting salary for M.S.W. social workers, often without stating exactly what is meant by the term. In statistical analyses, the "typical" represents an attempt to find a single number, or a series of numbers, that is most representative of a whole group of values. In a raw data set (e.g., Table 2.1 in Chapter 2), a typical case would be the one that is most representative of all cases in the data set.

CENTRAL TENDENCY

In statistical analyses, three terms are used to describe what is meant as "typical" within a data set—*mode, median,* and *mean.* These terms are not interchangeable and have specific meanings that differ in important ways. They must be used correctly to avoid confusion and to avoid misrepresenting what a data set looks like. These three terms are grouped under the general category of *central tendency*.

The three measures of central tendency have two basic uses:

1. *They summarize data.* They report one value (or number or score) that tells us something important about the characteristics of the distribution of a variable. For example, a social work agency may state in its annual report that it processed an average of 2.5 new client intakes a day during the previous year.
2. *They provide a common reference point for comparing two groups of data.* For example, an agency that hires both M.S.W.s and B.S.W.s as beginning social workers may report an average monthly starting salary of $2,100 for M.S.W.s and $1,700 for B.S.W.s. This one number, average monthly starting salary, helps communicate the agency's fiscal policy toward hiring beginning M.S.W. and B.S.W. social workers.

The Mode

The *mode* is the value in a distribution of values that occurs most frequently. In the 15 ages of clients presented in the array below, 42 is the mode because it occurs more frequently than any of the other values—in this case, four times more frequently.

Ages of clients ($N = 15$):
28, 31, 38, 39, 42, 42, 42, 42, 43, 47, 51, 54, 55, 56, 60

Sometimes in data sets more than one value will occur "most frequently." If we were to draw a histogram of the distribution of the values, it would have two distinct "peaks." If this situation occurs, we would report both values as the mode for the data set and would describe the distribution of the variable as *bimodal.* In the following example, the values in the bimodal array of years of prior social work experience among 22 social workers in a family service agency contain two modes—0 and 7. Both values occur five times each.

Years of prior social work experience ($N = 22$):
0, 0, 0, 0, 0, 1, 2, 2, 3, 4, 5, 5, 6, 7, 7, 7, 7, 7, 8, 9, 11, 14

When data are available in grouped form, the mode can be reported in one of two ways. We may simply report the grouping that had the largest frequency as the mode, or we can report the midpoint of the interval with the highest frequency. Table 3.1 portrays grouped job satisfaction scores of 50 social workers. For these data, the interval containing the largest frequency is 48–50, which includes the values 48, 49, and 50. This value range occurred seven times. The mode could be reported as 49, since the middle value for this interval is 49.

TABLE 3.1 Grouped cumulative frequency distribution: Job-satisfaction scores for social workers

Scores	Absolute Frequency	Cumulative Frequency (High—Low)	Cumulative Frequency (Low—High)
81–83	3	3	50
78–80	1	4	47
75–77	5	9	46
72–74	6	15	41
69–71	1	16	35
66–68	5	21	34
63–65	4	25	29
60–62	1	26	25
57–59	1	27	24
54–56	4	31	23
51–53	3	34	19
48–50	7	41	16
45–47	1	42	9
42–44	4	46	8
39–41	2	48	4
36–38	2	50	2

Of the three measures of central tendency, the mode is the most unrestricted—that is, it has the fewest requirements for its use. It can be used with all four measurement levels (nominal, ordinal, interval, and ratio). However, the mode is not used as often as the other measures of central tendency, as it lacks the precision of the other measures. When we have ordinal, interval, or ratio level data, we usually can obtain more accurate and representative descriptions by using one or both of the other two measures of central tendency. As can be seen in Table 3.2, the most common or frequent value of a distribution of scores is not necessarily the most accurate portrayal of a "typical" value. The mode is clearly not in the center of the distribution; rather, it is toward the high end of it (the 57-59 group).

The Median

If data can be formed into an array, that is, if they are at least at the ordinal level, the *median* can be used. The median divides an array of values into two identical halves. The example of Distribution A presents an array of 21 values for the variable *number of treatment sessions attended.* The median is 9 sessions because 9 coincides with the point that divides the 21 values into two equal parts. There are just as many values, or cases (10), above 9 as there are below 9.

Distribution A: Number of treatment sessions attended (*N* = 21)
2, 2, 2, 3, 3, 4, 5, 5, 7, 8, **9**, 10, 11, 11, 14, 14, 15, 16, 18, 29, 41

TABLE 3.2 Grouped cumulative frequency distribution: Job-satisfaction scores for social workers

Scores	Absolute Frequency	Cumulative Frequency (High−Low)	Cumulative Frequency (Low−High)
57–59	10	10	33
54–56	6	16	23
51–53	7	23	17
48–50	3	26	10
45–47	2	28	7
42–44	1	29	5
30–41	4	33	4

If there had been an even number of values, the median would be the average of the two most central values. In the array in Distribution B, 4.5 is the median.

Distribution B: Number of treatment sessions attended ($N = 24$)
1, 1, 1, 1, 1, 1, 2, 2, 3, 3, 3, **4, 5,** 6, 6, 7, 8, 11, 11, 13, 14, 15, 17, 20

The median (4.5) in Distribution B does not coincide with any specific value in the distribution. It cannot, because with an even number of cases there is no case that falls at exactly the midpoint. This observation underlines a point that is important to our understanding of the median. Contrary to a common misconception, the median is not synonymous with the value of the middle case in an array of data (although it sometimes works out that way, as in Distribution A).

Even when the array contains an odd number of cases, the median is likely to be a fraction that coincides with no actual case value. That is because the formula used to compute the median takes various conditions into consideration, such as case values with a frequency of zero and others with a frequency greater than one that occur near the center of the array. When these conditions exist, computation involves viewing numbers as intervals, as discussed in Chapter 2 regarding assignment of cases within grouped frequency distributions. Hand computation of the median can get pretty complicated; fortunately, most computer statistical software packages can compute a median in a matter of seconds once the raw data have been entered.

Of the three measures of central tendency, the median is affected the least by the presence of extremely atypical values. If we look at Distribution A, we see that one client had been seen many more times than any other—41 times. This atypical value is known as an *outlier.* If a histogram were created for the variable, this atypical value would lie outside the area where most of the other values are found (in Distribution A, between 2 and 29). Because the median coincides with the midpoint of the values in an array, the one client who was seen 41 times does not distort the median. This client is "canceled out" by the first client, who was seen 2 times and is the counterpart at the extreme other end of the array. The client seen 29 times similarly was canceled out by the second client seen 2 times, and so on.

The Mean

When a raw data set is at the interval or ratio level of measurement, the *mean* can be used when trying to find its typical value. It is the most easily understood, the best known, and the most useful of the three measures of central tendency. It is nothing more than the sum of all the values in a distribution divided by the total number of values—what we refer to in everyday speech as the "average." We can express this as a simple formula:

$$\text{Mean} = \frac{\text{Sum of all values in a distribution}}{\text{Total number of values in the distribution}}$$

Since a mean can be computed for any set of numbers, we might ask when it is not correct to report a mean. A mean should be reported only for those data that are at the interval or ratio levels of measurement. As we noted in Chapter 1, values expressed as numbers do not guarantee that a variable is at least at the interval level. For example, many newly constructed instruments for measuring attitudes generate a numerical score for each case, yet the data are not sufficiently precise to justify treating them as at the interval or ratio level. In these situations the median and/or the mode should be reported.

It is likewise not appropriate to report the mean rankings for ordinal-level data. Despite the fact that rankings are expressed as numbers, they really reflect only ordinal measurement. A student may rank Number 3 in one class, Number 4 in another, and Number 2 in a third class. While a mean ranking could mathematically be obtained (the student's "average" rank would be 3), to report it could be misleading. The reader of a report is right to assume that a mean was computed using interval or ratio level data.

Another issue must be considered when deciding if it is appropriate to use the mean. Unlike the median, the mean uses all the values within a distribution in its computation, not just some of the more central scores in an array previously formed. This characteristic can promote accuracy or distortion, depending on the absence or presence of outliers. Even one or two outliers can easily distort the mean if the total number of cases is small. With larger samples a few outliers will cause less distortion. If the presence of outliers is likely to produce a mean that is either too large or too small to be considered "typical" of the data, the median and/or the mode should be reported in its place.

THE MODE, THE MEDIAN, OR THE MEAN?

The answer to the question of which measure or measures of central tendency to use when describing a distribution of values is not always easy. With nominal level variables, the mode is all that can be used. However, if data are at the interval or ratio level, the final decision is often more one of ethics than of rules. Like frequency distributions and graphs, we want to use central tendency to provide our readers with a mental image—a shorthand description of what the data look like. Yet in some situations no single measurement of central tendency would accurately represent the data. An example using data for closed cases from a human service agency will help illustrate this point.

Example

Suppose an agency administrator wishes to see if the agency really is using short-term crisis intervention treatment, as stated in the agency's mission statement. Data are collected and analyzed for cases closed during the month of December (Table 3.3). Figure 3.1 is a frequency polygon that presents these data (from Table 3.3) with the mode, median, and mean indicated.

The variable *number of interviews* is at the ratio level of measurement, and there are a large number of cases ($N = 1,290$). These conditions suggest that the mean might be the best central tendency measure to report, but what about outliers? The shape of the polygon indicates that the presence of outliers may produce a distorted mean.

The mean number of interviews in Table 3.3 is actually 4.37, primarily because of the fairly large number of cases (outliers) that were seen ten times ($n = 35$). It is apparent from Figure 3.1 that a client interviewed four or five times (rounding down or up) would not really be "typical" of clients within the

TABLE 3.3 Frequency distribution: Number of interviews for cases closed at XYZ agency during December

Number of Interviews (a)	Number of Cases (b)	Total Number of Interviews (Column a) (Column b)
1	55	55
2	35	70
3	55	165
4	40	160
5	25	125
6	15	90
7	10	70
8	20	160
9	5	45
10	35	350
Totals	295	1,290

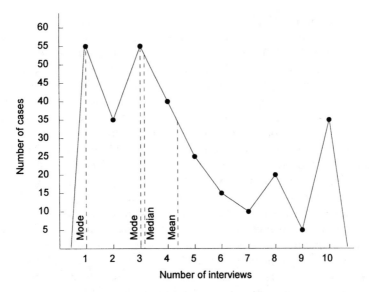

FIGURE 3.1 Frequency polygon: Number of interviews for cases closed at XYZ agency during December (from Table 3.3)

agency. There are actually four other values (1, 2, 3, and 10) that occurred as frequently as or more frequently than either 4 or 5. Because three of the values (1, 2, and 3) are all clustered well to the left of the mean, it appears to be too high to reflect what is a typical value for the distribution.

If the mean would not be a representative portrayal of the data, what about the median? It falls between the values 3 and 4 and very close to one of the most common values (3). As a choice for a single measure of central tendency to represent the data, it is fairly good. However, it does not even hint that a fairly sizable group of clients ($n = 35$) were interviewed 10 times, a fact that may be a little surprising and possibly valuable for an agency that is generally believed to engage in short-term crisis intervention. It also does not confirm the more predictable finding that a large number of clients ($n = 55$) were interviewed only once. In short, the median may be better than either the mean or the mode for presenting what is typical, but it is far from perfect for reporting central tendency for these data.

The distribution is bimodal, with the two modes falling at one and three interviews. But if we were to use only the mode, we would be suggesting that either one or three interviews (both small numbers) is typical of a client when, in fact, fewer than half of all the clients were seen four times or less. As with the median, the mode alone provides no hint of the possibility that a sizable percentage of "crisis intervention cases" were interviewed a fairly large number of times. Besides, the mode is really most appropriate for nominal level data. It treats different values of a variable as if they were differences of kind only. It would not take into account that the values in the example reflect quantitative differences, a fact that is critical to an accurate interpretation of the data.

In this example, as in most data sets, no single measure of central tendency would adequately summarize what the data looked like. Any one measure would have the potential to mislead. Yet all three—the mode, the median, and the mean—contribute something toward visualizing how the data looked. The fact that the data were bimodal, with modes at one and three interviews, indicates that short-term treatment occurs quite often within the agency. Yet the median best reflects what is typical. It uses some of the available precision of measurement, more than the mode but less than the mean. It at least suggests that short-term treatment may not be as typical of the agency as one might believe from reading the agency's mission statement. If the mean is presented, the fact that it is over four is even stronger evidence that a sizable number of clients clearly are not the recipients of crisis intervention services.

When in doubt about which measure of central tendency to use, report all three. If the mode, median, and mean are all reported, an experienced reader of research reports, or even one with just a good understanding of central tendency, will be able to compare them and piece together a reasonably good picture of how the data were distributed. Any one measure alone might mislead; all three taken together would present an accurate communication—which, as we have repeatedly emphasized, is the goal of descriptive statistics.

VARIABILITY

While one or more measures of central tendency can tell us much about the distribution of the values of a variable, they fall far short of giving us a complete picture. We still must ask: Did most values tend to cluster around the typical value, or, if not, how widely did they vary from it? To get an accurate description of the distribution of the values of a variable, we need to add another summary description—one that gives us an indicator of the degree of variation that occurred. We must add a measure of *variability,* also called a measure of dispersion.

Let us see why this is important. Suppose that there are two class sections of a graduate social work research seminar. The ages of the 15 students in each section are as follows:

Ages: Section 1
21, 22, 24, 24, 26, 29, 30, 31, 32, 33, 36, 38, 38, 40, 41

Ages: Section 2
27, 28, 28, 29, 29, 30, 30, 31, 32, 32, 33, 33, 34, 34, 35

If we were to report the measures of central tendency for the variable *age* for these two sections, they would be identical. The mean (31) and median (31) are the same for both distributions. Only the mode would give any hint that the distribution of ages in the two sections is different, and they are quite different. If a 23-year-old student was to register late for the seminar, the student might feel more comfortable in Section 1 than in Section 2. However, a student who is 39 years of age might feel more comfortable in Section 2, but neither student would know this if only central tendency data were available. A measure of variability would help to complete the picture.

If data are at the nominal level of measurement, variability can be communicated only in a frequency distribution or graph. If they are at the ordinal level of measurement (i.e., their values can be rank ordered and placed in an array), some measures of variability can be computed, but they communicate little about how the values are dispersed. So, they are rarely used. If a variable is at the interval or ratio level of measurement, however, measures of variability are very helpful. The five measures of variability that we will discuss are: (a) range, (b) interquartile range, (c) mean deviation, (d) variance, and (e) standard deviation.

The Range

The *range* is the distance encompassed by the difference between the *maximum value* (largest) and the *minimum value* (smallest) in an array of values. Expressed as a formula, the range is:

$$\text{Range} = \text{maximum value} - \text{minimum value} + 1$$

The formula differs slightly from the way we use the word *range* in common English usage. Why is the range not simply the difference between the maximum value and the minimum value? We add 1 to the difference so that the range reflects the total *number* of values of the variable that it encompasses. For example, in a distribution of the variable *age,* with a maximum age of 35 and a minimum age of 30, the range is 6 (35 − 30 + 1 = 6). There are potentially 6 different ages that are included within the range: 35, 34, 33, 32, 31, and 30. The range for Section 2 in the previous example is 9 (35 − 27 + 1 = 9). There are potentially 9 different ages included in the data: 27, 28, 29, 30, 31, 32, 33, 34, and 35. The range for Section 1 is 21 (41 − 21 + 1 = 21). The larger range in Section 1, compared with that of Section 2, indicates a greater variation in the students' ages.

The range is a useful measure of variation in that it can be computed quickly and easily. Calculating the range also suggests the number of intervals to employ in creating grouped frequency distributions as well as the most appropriate interval size. However, like the mean, it is easily distorted by the presence of outliers. One outlier at either end of an array can greatly increase the range of a data distribution and suggest much more variability within the data set than is actually present. Let us suppose that the 41-year-old student in Section 1 drops out of the research seminar and is replaced by a 64-year-old student. The distribution of values would now look like this:

Ages: Section 1
21, 22, 24, 24, 26, 29, 30, 31, 32, 33, 36, 38, 38, 40, 64

Even though only one value has changed, the range for Section 1 has jumped from 21 to 44 (64 – 21 + 1 = 44). A student considering enrolling in Section 1, upon learning that the range for the students' ages is 44 years, would mistakenly assume that the ages of the students are much more diverse than they really are.

The vulnerability of the range to the influence of outliers is an undesirable characteristic, especially when comparing the ranges of two distributions of the same variable. The presence of outliers in one distribution and not in the other can give a very misleading impression about the degree of similarity of the two distributions.

The Interquartile Range

One way to handle the problem of outliers would be to use another measure of variability rather than the range. Instead of using the maximum and minimum values to obtain a range, we could report variability as the distance between the 75th and 25th percentiles. This distance is known as the *interquartile range.*

We will recall that the median falls at the 50th percentile, or midpoint (also known as the *second quartile*), of an array of data where half the values fall above it and half below. In such an array, the 25th percentile (also known as the *first quartile*) would fall at the point where one-fourth of values in the array would have lower or smaller values. The 75th percentile (also known as the *third quartile*) would fall at the point where three-fourths of values in the array would have lower or smaller values.

Once an array has been formed and the 75th and 25th percentiles have been determined, the interquartile range is found by subtracting the 25th percentile from the 75th percentile. The interquartile range is a more stable measure than the range as a measure of variability for the same reason that the median is more stable than the mean as a measure of central tendency. Outliers cannot distort it in the same way they would distort the range because only their positions in the array, not their actual case values, are used in the final computation of the interquartile range.

Like the formula for the median, the formula for the interquartile range can be a little difficult to compute by hand. It is often a fraction, as it would be with any of the examples of distributions that we have previously used. With computer analysis of data this presents few problems, as its computation takes only a few seconds once the data have been entered.

The Mean Deviation

The range and the interquartile range have advantages and are useful in certain situations, but both are a little "crude" as measures of variability. This is because they do not use every case value in their final calculations. The *mean*

TABLE 3.4 Deviations from the mean

Value	−	Mean	=	Deviations from the Mean
1	−	3	=	−2
2	−	3	=	−1
3	−	3	=	0
4	−	3	=	1
5	−	3	=	2
		Total		0

deviation is one of three measures of variability that we will discuss that is derived from all the values in a distribution. It is the average amount that the values of a variable differ (or deviate) from the mean within a distribution. Like other measures of variability, it indicates only the amount of variation among values of a variable, not their absolute values.

Table 3.4 lists 5 values (1, 2, 3, 4, 5), their mean (3), and the deviation score of each (−2, −1, 0, 1, 2), which is the difference between each respective value (1, 2, 3, 4, 5) and their mean (3).

The formula for the mean deviation is:

$$\text{Mean deviation} = \frac{\text{Sum of deviation values (ignoring sign)}}{\text{Number of cases}}$$

To compute the mean deviation for the data in Table 3.4, we would proceed as follows:

$$\text{Mean deviation} = \frac{2 + 1 + 0 + 1 + 2}{5}$$
$$= 1.2$$

The mean deviation is relatively easy to compute and interpret, but it is rarely reported in our professional literature. Once computed, we use it to produce the variance and the standard deviation, the two most widely used measures of variability.

The Variance

Obtaining the *variance* requires subtracting the mean of the distribution from each value, squaring each difference, and then dividing the sum of the squared differences by the total number of values.

To compute the variance for the data in Table 3.4, we would proceed as follows:

$$\text{Variance} = \frac{\text{Sum of squared deviations from the mean}}{\text{Number of cases}}$$

$$\text{Variance} = \frac{(-2)^2 + (-1)^2 + 0^2 + (1)^2 + (2)^2}{5}$$

$$= \frac{4 + 1 + 0 + 1 + 4}{5}$$

$$= \frac{10}{5}$$

$$= 2$$

The variance sometimes is reported as a descriptive statistic within our professional literature. It also is used in some of the statistical computations that we will discuss later, such as those in Chapters 10 and 12.

The Standard Deviation

The square root of the variance—the *standard deviation*—appears frequently in quantitatively oriented reports. It is a favorite for describing the variability of data, and it is used in many other types of statistical analyses as well.

The standard deviation requires interval or ratio level data. It also is most appropriately used with fairly large samples and with variables that, if graphed, would produce a frequency polygon that has a particular shape—the normal distribution (to be discussed in Chapter 4).

Like the mean deviation and the variance, the standard deviation uses all values in its computation. It tells us to what degree the values cluster around the mean, which makes it extremely useful. As we shall see in Chapter 4, when reported with the mean in appropriate situations, it allows us to determine where a given value falls relative to other values and to reconstruct the distribution of all the values of a variable. For the moment, we will concentrate on how the standard deviation is computed from those values.

Computing the standard deviation involves eight steps, many of which we have already discussed in relation to other measures of variability. (The columns refer to Table 3.5.)

1. List the values in a distribution in column *a*.
2. Obtain a mean of the values in column *a*.
3. List the mean in column *b*.
4. Subtract the mean from each value in column *a*, and place this value in column *c*.
5. Square each value in column *c*, and place this value in column *d*.
6. Add column *d*.
7. Divide the sum of column *d* by the total number of values in column *a*.

8. Obtain the square root of the number computed in Step 7 (the variance). This number is the standard deviation of the values in column *a*.

The data in the distribution below give the number of years of employment of the six social workers who work in agency A:

Years of employment: Agency A
5, 5, 6, 6, 7, 7

Using the data, we can compute the standard deviation, as presented in Table 3.5.

Now suppose that agency B also has six social workers, but their years of employment are as follows:

Years of employment: Agency B
1, 2, 4, 8, 10, 11

Again, we can compute the standard deviation, as presented in Table 3.6.

The distribution of years of employment in agency B shows more variation than does the distribution for agency A. This is also reflected in their respective standard deviations—3.87 for agency B and .82 for agency A. This demonstrates an important point: In comparing two distributions of measurements of the same variable, larger standard deviations reflect more variation, and vice versa.

A pictorial representation of the above data will further demonstrate the meaning of standard deviation. Figures 3.2 and 3.3 each show how the data on length of employment in the two agencies would look if displayed as weights on a scale.

TABLE 3.5 Determining the standard deviation of years of employment for agency A

Step 1 (a) Value	−	Step 3 (b) Mean	=	Step 4 (c) Deviations from Mean	Step 5 (d) Squared Differences from Mean
5	−	6	=	− 1	1
5	−	6	=	− 1	1
6	−	6	=	0	0
6	−	6	=	0	0
7	−	6	=	1	1
7	−	6	=	1	1
		Total		0̄	Step 6 = 4̄

$$\text{Step 7} = \frac{4}{6}$$
$$= .67 \quad \text{(Variance)}$$

$$\text{Step 8} = \sqrt{.67}$$
$$= .82 \quad \text{(Standard deviation)}$$

TABLE 3.6 Determining the standard deviation of years of employment for agency B

Step 1 (a) Value	−	Step 3 (b) Mean	=	Step 4 (c) Deviations from Mean	Step 5 (d) Squared Differences from Mean
1	−	6	=	− 5	25
2	−	6	=	− 4	16
4	−	6	=	− 2	4
8	−	6	=	2	4
10	−	6	=	4	16
11	−	6	=	5	25

$$\text{Step 6} = 90$$

$$\text{Step 7} = \frac{90}{6}$$
$$= 15 \quad \text{(Variance)}$$

$$\text{Step 8} = \sqrt{15}$$
$$= 3.87 \quad \text{(Standard deviation)}$$

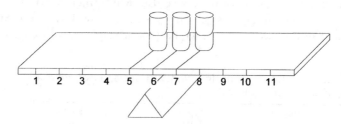

FIGURE 3.2 Variability of years of employment for agency A (from Table 3.5)

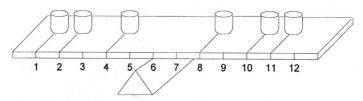

FIGURE 3.3 Variability of years of employment for agency B (from Table 3.6)

Note that Figures 3.2 and 3.3 both "balance" on 6 years, reflecting the fact that both distributions have a mean of 6 (years of employment). However, notice how much more variation is shown in agency B (Figure 3.3) than in agency A (Figure 3.2), which is reflected in agency B's larger standard deviation.

CONCLUDING THOUGHTS

The standard deviation represents the distance between the mean and a certain point on a histogram (frequency polygon), reflecting the distribution of a variable. A comparison of two standard deviations computed from their respective distributions indicates which distribution has greater spread between that point and the mean (the one with the larger standard deviation).

The mean is usually reported along with the standard deviation. This enables the reader of a research report to get a better picture of what the actual measurements of a variable within a given data set look like.

To provide an even more complete image of the distribution of the measurements of a variable, researchers sometimes report what is referred to as a *five-number summary*. It consists of the minimum value, the first quartile (25th percentile), the median, the third quartile (75th percentile), and the maximum value. A graph known as a *Tukey boxplot* is also sometimes used to display the central tendency and dispersion of the distribution of a variable.

Adding or subtracting a fixed amount to all values of a distribution will affect the mean by increasing or decreasing it by that amount. It will not affect the standard deviation for that distribution, since standard deviation reflects variation, and the amount of variation would not change. For example, when every employee of a human service agency receives a $2,000 raise, the mean salary will increase by $2,000, but the amount of variation in salary among employees (its standard deviation) will remain exactly the same. Similarly, if we were to repeat our measurement of years of employment one year later with the same employees in either agency, the average number would now be 7 years, but the respective standard deviations would be the same as they were a year earlier.

STUDY QUESTIONS

1. What does a frequency polygon look like when the distribution is described as bimodal? Provide an original example in your discussion.
2. Why is the median more likely to be an accurate description of interval level data than the mode? Explain.
3. Discuss why the median is a more stable measure of variability than the mean. Provide an original example in your discussion.
4. Why do we consider the mean to be a more precise measurement of central tendency than either the median or the mode?
5. Why may only one measure of central tendency be an inadequate description of a data set? Provide an original example in your discussion.

6. Discuss why the range is an especially unstable measure of variability and when the interquartile range may be preferable.
7. Which measures of variability discussed in this chapter consider all values of a variable in their computations? Why is this better than using only some of the values in a distribution? Explain.
8. How would an outlier tend to distort the mean deviation for a group of data?
9. How would a data set containing the values of a variable with a mean of 50 and a standard deviation of 3 compare with another data set containing the same variable but a mean of 50 and a standard deviation of 12?
10. How would adding the number 10 to each of the values of a variable affect its mean and standard deviation? Provide an example.
11. Fifteen students are registered in Section 1 and 15 in Section 2 of a research course. They took the same midterm exam, and their grades were distributed as follows:

    Section 1: 89, 56, 45, 78, 98, 45, 55, 77, 88, 99, 98, 97, 54, 34, 94
    Section 2: 77, 88, 87, 67, 98, 87, 55, 77, 45, 44, 88, 99, 69, 67, 98

    a. Calculate the mode, median, mean, range, variance, and standard deviation for both sections.
    b. Which section did better overall on the exam, Section 1 or Section 2? Fully justify your answer using the concepts presented in this chapter.
12. Locate an article in a social work professional journal that uses one or more of the measures of central tendency and variability presented in this chapter. Answer the following questions in relation to the article.
    a. Do you feel the author reported the data accurately when referring to the data set used in the research study? Why or why not?
    b. What other measures of central tendency and variability would you have liked the author to report? Why would they have been helpful?
13. Discuss why a mean should not be used with ordinal level variables.
14. For each of the following, indicate the measure of central tendency that would be most appropriate, and indicate why.
    a. An income distribution in which 97 percent of the cases are in a range of $20,000 to $50,000 and a few cases are between 0 and $5,000.
    b. Data reporting the religious preferences of 100 social work students.
    c. A grouped frequency distribution of the variable *age* that has an open-ended interval of "over 65 years of age."
15. The measures of central tendency have been reported for three different social service agencies for the variable *number of years employed in the agency* as follows:

    Agency A: mode 16, median 17, mean 16
    Agency B: mode 4, median 7, mean 10
    Agency C: mode 1, median 3, mean 6

    Use the measures of central tendency to describe and compare the staff of the three agencies.

chapter 4

Normal Distributions

Chapter 2 presented ways that tables and graphs can be used to portray the distributions of values of a variable within a data set. Chapter 3 illustrated the various methods that can be used to summarize two important characteristics of a distribution of values—central tendency and variability. This chapter will demonstrate how the shape of a distribution (the polygon that can be formed from a histogram), along with measures of its central tendency and variability, can be used to present a more complete description of the distribution of a variable.

SKEWED DISTRIBUTIONS

In Chapter 2 we showed how a frequency distribution can be graphically portrayed as a frequency polygon. The shape of a frequency polygon reflects where the various measures (values) of a variable tend to cluster. Some polygons reflect the fact that relatively large numbers of values cluster at the left side where lower measurements of the variable are displayed; others reflect the opposite pattern.

Suppose, for example, that a hospital administrator wished to study changes in admission diagnoses over a six-year period. She might wish to substantiate her impression that the hospital is experiencing a decline in some diagnoses and an increase in others. The data might look like those in Table 4.1 for a diagnosis such as emphysema.

Just by glancing at the data in Table 4.1, it is easy to see that the number of emphysema patients admitted to the hospital over the six-year period has declined

TABLE 4.1 Cumulative frequency
distribution (*N* = 210): Emphysema
patients admitted to XYZ hospital by year

Year	Absolute Frequency	Cumulative Frequency
1990	60	60
1991	50	110
1992	40	150
1993	30	180
1994	20	200
1995	10	210

over time. This trend is even more apparent when the data are placed in a histogram such as the one in Figure 4.1.

The midpoints of the bars in the histogram in Figure 4.1 are connected. The
line joining the midpoints to form a frequency polygon is called a *curve*. Distributions like the one shown in Table 4.1 and reflected in the frequency polygon
in Figure 4.1 are referred to as *skewed*. A skewed distribution is not symmetrical—that is, its ends do not taper off in a similar manner in both directions. Note

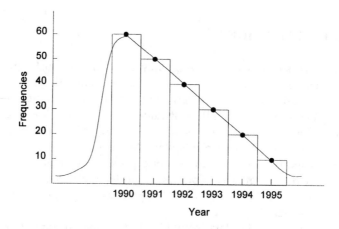

FIGURE 4.1 Positively skewed frequency
polygon (from Table 4.1)

TABLE 4.2 Cumulative frequency distribution (N = 210): HIV-positive patients admitted to XYZ hospital by year

Year	Absolute Frequency	Cumulative Frequency
1990	10	10
1991	20	30
1992	30	60
1993	40	100
1994	50	150
1995	60	210

that the frequency polygon in Figure 4.1 has a "tail" on the right side. A curve like the one in Figure 4.1, which is skewed to the right, is called a *positively skewed* distribution.

Trends in admissions of HIV-positive cases over the six-year period in the same hospital might reflect a very different pattern from those of emphysema admissions. Table 4.2 and Figure 4.2 illustrate this.

The distribution in Figure 4.2 is also skewed, but this time the tail of the frequency distribution is to the left. A curve that is skewed to the left is called a *negatively skewed* distribution.

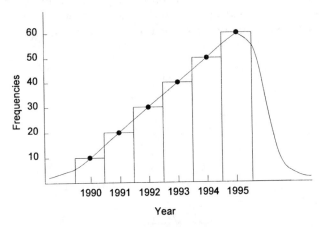

FIGURE 4.2 Negatively skewed frequency polygon (from Table 4.2): HIV-positive patients admitted to XYZ hospital by year

FIGURE 4.3 The normal curve

NORMAL DISTRIBUTIONS

A distribution that is free of skewness is symmetrical, as is the frequency polygon used to display it. Among the various symmetrical distributions of values of variables that can occur, one special form is called a *normal distribution*. Many variables that are of interest to social work researchers tend naturally to form normal distributions. A normal distribution is "bell-shaped" when portrayed using a frequency polygon; the curve thus formed is referred to as the *normal curve*. One example of a normal curve reflecting a normal distribution is presented in Figure 4.3.

Distributions of all variables that tend to be normally distributed within any given population share the same properties. What are the other properties of a normal curve, in addition to being symmetrical and bell-shaped? For one thing, in a normal curve the mode, median, and mean all occur at the highest point and in the center of the distribution, as in Figure 4.4. Note that in skewed curves the mode, median, and mean occur at different points, as in Figure 4.5a and b.

The ends of the normal curve extend toward infinity—they approach the horizontal axis (*x*-axis) but never quite touch it. This property represents the

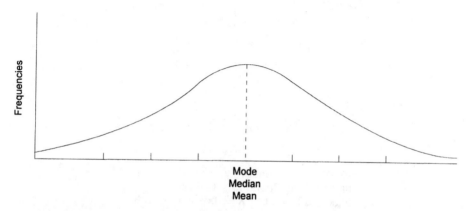

FIGURE 4.4 The normal distribution

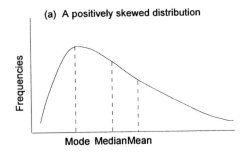

(a) A positively skewed distribution

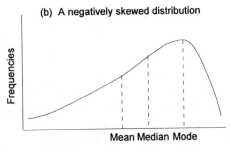

(b) A negatively skewed distribution

FIGURE 4.5 Skewed distributions

possibility that, while the curve contains virtually all measurements of the variable within a population, a small number of values may exist that reflect extremely large or extremely small measurements of the variable (outliers). It also reflects the fact that, at a higher level of abstraction, a total universe or population is never static because it is always subject to change as cases are added or deleted over time. Populations are always evolving.

The horizontal axis that lies below the normal curve can be divided into six equal units—three between the mean and the place where the curve approaches the axis on the left side, and three between the mean and the place where it approaches the axis on the right. These six units reflect the amount of variation that exists within virtually all values of a normally distributed variable. The larger the size of the units, the flatter the curve will be and the more variation it will reflect. How large should the units be for portraying the distribution of any given normally distributed variable? Their exact size is determined by using the standard deviation formula that was presented in Chapter 3.

The number thus produced (the width of each unit along the x-axis) is referred to as the standard deviation for the distribution of a given variable. Figure 4.6 shows a normal curve with measurements reflecting three units of variability (standard deviations) to the left of the mean and three units of variability to the right of the mean. Note that the units are labeled to reflect the

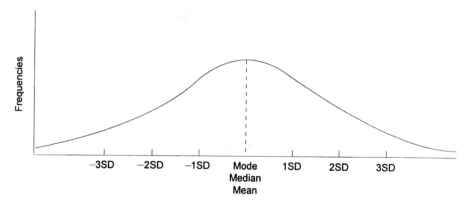

FIGURE 4.6 The normal distribution

number of standard deviations (SD) that each falls from the mean. Units to the left of the mean (where values are smaller) use the minus sign (-1SD, -2SD, and -3SD), and units to the right of the mean (where values are larger) use no sign at all, which means they are positive (1SD, 2SD, and 3SD).

The term *standard deviation* can be a little misleading. What is "standard" about standard deviation? First, once computed for a distribution of values of a given variable, it does not change. Also, "standard" means that, in a normal distribution of any variable, virtually all values (except for .26 percent) will fall within + or - three standard deviations from the mean. However, the sizes of standard deviations vary from data set to data set or from variable to variable, based upon the measurements that are used in their computation.

Different normal curves, therefore, tend to have different means and different standard deviations. Figure 4.7a, b, and c demonstrates how this occurs by comparing three pairs of normal curves. It also demonstrates the fact that normal curves show variety; that is, they may be high and narrow, low and wide, or anything in between. They are usually drawn to suggest the degree of variation present in the distribution of a variable that is, the size of its standard deviation. Flatter curves suggest relatively large standard deviations, and taller ones reflect relatively small standard deviations.

The total area underneath a normal curve is equal to about 100 percent of the curve (99.74 percent, to be exact). Parts of the curve are represented as percentages of the total area of the curve. Not surprisingly, 50 percent of the values of a normal curve fall below the mean, and 50 percent fall above the mean. Mathematicians also have figured out what proportions of the normal curve fall within its various segments. Figure 4.8 illustrates these proportions.

By looking at Figure 4.8, we can see that the area of a normal curve between a point on the horizontal axis (e.g., -2SD) and the mean is equivalent to the area of the curve between the comparable point on the other side of the mean (e.g., 2SD) and the mean. This makes sense because, as we have already noted, a normal curve is symmetrical. If we were to add all the percentages within each of the

(a) Equal means, unequal standard deviations

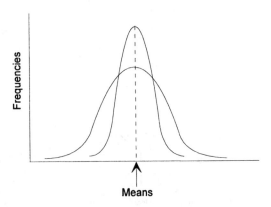

(b) Unequal means, equal standard deviations

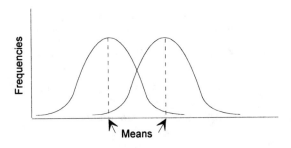

(c) Unequal means, unequal standard deviations

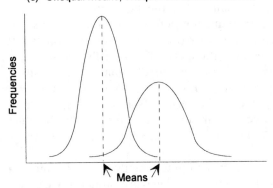

FIGURE 4.7 Variations in normal distributions

segments of the frequency polygon, they would equal 99.74 percent of the curve. We could also add the respective segments of the normal curve to learn that 47.72 percent of it (34.13% + 13.59% = 47.72%) falls between the mean and 2SD *above* it and also between the mean and 2SD *below* it, or that 68.26 percent (34.13% + 34.13% = 68.26%) falls within 1SD of the mean.

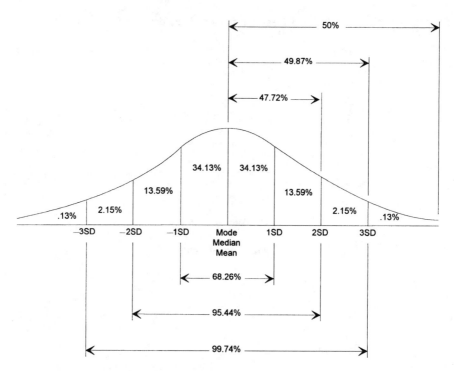

FIGURE 4.8 Proportions of the normal curve

Now let us look at Figure 4.8 from a different perspective. Up to this point, we have viewed the numbers in the figure as areas or portions of a normally distributed curve, or frequency polygon, but they are also something else. They are the percentage of values (persons, cases, or objects) that fall within the respective distances from the mean of a normally distributed variable. If, for example, Figure 4.8 were a frequency distribution of a normally distributed variable, such as *height of female social work students,* the figure would tell us that the height of 47.72 percent of all female social work students (34.13% + 13.59% = 47.72%) falls between the mean and 2SD from it; the height of 68.26 percent of them (34.13% + 34.13% = 68.26%) falls between +1SD and –1SD from the mean, and so on.

If we knew or could compute exactly what the mean and standard deviation for the height of female social work students were, we could then know even more. We could assign actual heights to the mean and standard deviation points in Figure 4.8 and make statements like "68.26 percent of the heights of female social work students fall between _____ inches tall and _____ inches tall." Understanding that the percentage of the area under a normal curve is also the percentage of values that fall within a certain area of distribution of a normally distributed variable is *critical* to understanding the material in this chapter and other parts of our discussion of statistics.

CONVERTING RAW SCORES TO *Z* SCORES

When we encounter values of a variable based on measurements taken from two different populations, we are sometimes unable to make direct comparisons between them. For example, we could not compare the class standing of two individuals who attend two different colleges directly knowing only their grade point averages unless the distribution of grades in the two colleges is identical. Such a situation would be very unlikely. A meaningful comparison would be possible only if we knew how the grade point average of each student compares with that of other students in his or her respective college. In other words, we would have to convert the two grade point averages to a common standard.

It is possible to use a common standard to compare values of a variable taken from two different populations if the variable is normally distributed within both populations. We do this through the use of *z* scores. *Z* scores are raw scores converted to standard deviation units. Every raw score in any normal distribution can be given an equivalent *z* score that reflects how many standard deviation units it falls from the mean. The relative positions of two *z* scores taken from two different normal distributions can then be compared directly with one another.

Because of the nature of the normal curve, *z* scores can be converted to *percentiles,* an even more familiar term. For example, suppose Axel's score on a midterm statistics exam was 75. Using *z* scores we might determine that such a score would fall at the 82nd percentile within his course section. Durshka's score on a statistics exam in a different midterm section was also 75, but her score might have fallen at the 92nd percentile among students in her section. Durshka can be assumed to have done better than Axel on her midterm exam, at least in one sense.

To convert a raw score into a *z* score, the following formula is used:

$$Z \text{ score} = \frac{\text{Raw score} - \text{mean}}{\text{Standard deviation}}$$

A *z* score reflects the number of standard deviation units that a given raw score falls from the mean of the distribution that contains the score. Thus, the mean itself would have a *z* score of 0.00. A *z* score of 1.00 is one standard deviation above the mean, a *z* score of 2.00 is two standard deviations above the mean, and so on. A *z* score is either positive or negative, depending upon whether its raw score is larger or smaller than the mean. A positive *z* score indicates a raw score above the mean, and a negative *z* score indicates a raw score below the mean.

As long as we know the mean and the standard deviation of a distribution from which any raw score is obtained, we can compute its *z* score. *Z* scores are often not whole numbers, but are fractions of numbers, such as $z = 2.11$ or $z = -2.24$. A figure such as Figure 4.8 would not be able to tell us the percentage of values (or the area under the normal curve) that would fall between a fractional *z* score and the mean of the frequency distribution that contains its raw score. We would have to go to a table such as Table 4.3 to assist us in converting our fractional z scores into percentiles. Table 4.3 tells us the area of a normal curve (and the corresponding percent of values) that fall between a whole or fractional z score and the mean within any normal distribution. Note that the number alongside 1.0 in the lefthand column is 34.13, the area of the normal curve between the mean and either +1SD or -1SD (see Figure 4.8). Also, the number alongside 2.0 in the lefthand column is 47.72, the sum of the numbers 34.13 and 13.59 in Figure 4.8. The 47.72 represents the percentage of values in any normal distribution that fall between the mean and either 2SD or -2SD.

In Table 4.3, the whole number and the first decimal of a z score are found in the lefthand column. The second decimal is seen in the column headings that run across the top of the table. The area of the normal curve between a given z score (obtained by using the z score formula) and the mean would be the number in the body of the table where the appropriate line and column intersect. For example, to find the area of the curve between a raw score and the mean when the raw score's z score computes to 1.55, we would first go down the lefthand column in Table 4.3 to 1.5. Then we would move right across the table to the .05 column (to pick up the second decimal). The number 43.94 appears at the intersection of the 1.5 line and the .05 column. That means that the area of the curve between our raw score and the mean would be 43.94 or, viewing it another way, that nearly 44 percent of all values (or cases) fall between that raw score and the mean.

For positive z scores (those to the right of the mean) we would *add* the area of the curve found in the body of Table 4.3 to 50.00 (corresponding to the area of the curve below the mean) to find the percentile where the raw score fell. In our example (using a z score of 1.55), we would add 43.94 to 50.00 to get 93.94. The raw score corresponding to a z score of 1.55 would fall at approximately the 94th percentile. It is logical to add 50.00 to the number found in Table 4.3 since we know that the raw score fell above the mean. All scores below the mean (50 percent of them in a normal distribution), plus the other 43.94 percent, fell below the raw score.

For negative z scores (those to the left of the mean), we would *subtract* the area of the curve found in the body of Table 4.3 from 50.00 (the percentile of the mean). If the z score in our example had turned out to be -1.55, we would subtract 43.94 from 50.00 to get 6.06. The raw score corresponding to a z score of -1.55 would fall at approximately the 6th percentile.

Table 4.4 provides additional examples of z scores and their corresponding areas and percentiles as obtained using Table 4.3.

Z scores are used appropriately with variables that form normal distributions within the population or approximate the normal curve. When the distribution is skewed, the area between -1SD and the mean is not equal to the area between 1SD and the mean. When distributions are skewed, a z score cannot be used to produce a standardized proportion of the distribution from which it was computed. For example, the distribution in Figure 4.9 is positively skewed.

TABLE 4.3 Areas of the normal curve

Area Under the Normal Curve Between Mean and z Score

z	.00	.01	.02	.03	.04	.05	.06	.07	.08	.09
0.0	00.00	00.40	00.80	01.20	01.60	01.99	02.39	02.79	03.19	03.59
0.1	03.98	04.38	04.78	05.17	05.57	05.96	06.36	06.75	07.14	07.53
0.2	07.93	08.32	08.71	09.10	09.48	09.87	10.26	10.64	11.03	11.41
0.3	11.79	12.17	12.55	12.93	13.31	13.68	14.06	14.43	14.80	15.17
0.4	15.54	15.91	16.28	16.64	17.00	17.36	17.72	18.08	18.44	18.79
0.5	19.15	19.50	19.85	20.19	20.54	20.88	21.23	21.57	21.90	22.24
0.6	22.57	22.91	23.24	23.57	23.89	24.22	24.54	24.86	25.17	25.49
0.7	25.80	26.11	26.42	26.73	27.04	27.34	27.64	27.94	28.23	28.52
0.8	28.81	29.10	29.39	29.67	29.95	30.23	30.51	30.78	31.06	31.33
0.9	31.59	31.86	32.12	32.38	32.64	32.90	33.15	33.40	33.65	33.89
1.0	34.13	34.38	34.61	34.85	35.08	35.31	35.54	35.77	35.99	36.21
1.1	36.43	36.65	36.86	37.08	37.29	37.49	37.70	37.90	38.10	38.30
1.2	38.49	38.69	38.88	39.07	39.25	39.44	39.62	39.80	39.97	40.15
1.3	40.32	40.49	40.66	40.82	40.99	41.15	41.31	41.47	41.62	41.77
1.4	41.92	42.07	42.22	42.36	42.51	42.65	42.79	42.92	43.06	43.19
1.5	43.32	43.45	43.57	43.70	43.83	43.94	44.06	44.18	44.29	44.41
1.6	44.52	44.63	44.74	44.84	44.95	45.05	45.15	45.25	45.35	45.45
1.7	45.54	45.64	45.73	45.82	45.91	45.99	46.08	46.16	46.25	46.33
1.8	46.41	46.49	46.56	46.64	46.71	46.78	46.86	46.93	46.99	47.06
1.9	47.13	47.19	47.26	47.32	47.38	47.44	47.50	47.56	47.61	47.67
2.0	47.72	47.78	47.83	47.88	47.93	47.98	48.03	48.08	48.12	48.17
2.1	48.21	48.26	48.30	48.34	48.38	48.42	48.46	48.50	48.54	48.57
2.2	48.61	48.64	48.68	48.71	48.75	48.78	48.81	48.84	48.87	48.90
2.3	48.93	48.96	48.98	49.01	49.04	49.06	49.09	49.11	49.13	49.16
2.4	49.18	49.20	49.22	49.25	49.27	49.29	49.31	49.32	49.34	49.36
2.5	49.38	49.40	49.41	49.43	49.45	49.46	49.48	49.49	49.51	49.52
2.6	49.53	49.55	49.56	49.57	49.59	49.60	49.61	49.62	49.63	49.64
2.7	49.65	49.66	49.67	49.68	49.69	49.70	49.71	49.72	49.73	49.74
2.8	49.74	49.75	49.76	49.77	49.77	49.78	49.79	49.79	49.80	49.81
2.9	49.81	49.82	49.82	49.83	49.84	49.84	49.85	49.85	49.86	49.86
3.0	49.87									
3.5	49.98									
4.0	49.997									
5.0	49.99997									

Source: The original data for Table 4.3 came from *Tables for Statisticians and Biometricians*, edited by K. Pearson, published by the Imperial College of Science and Technology, and are used here by permission of the Biometrika trustees. The adaptation of these data is taken from E. L. Lindquist, *A First Course in Statistics* (revised edition), with permission of the publisher, Houghton Mifflin Company.

TABLE 4.4 Examples of z scores and their corresponding areas and percentiles

z Score	Row	Column	Area Included Between Mean and z Score	Percentiles
.12	0.1	.02	04.78	54.78
1.78	1.7	.08	46.25	96.25
−2.90	2.9	.00	49.81	.19
1.15	1.1	.05	37.49	87.49
−1.15	1.1	.05	37.49	12.51

Area A is not equal to Area B, even though each area corresponds to 1SD from the mean.

Practical Uses of Z Scores

When used with normally distributed distributions, z scores make it possible for us to take any single interval or ratio value (raw score) from a normally distributed distribution and gain an accurate understanding of where it falls relative to the other scores by the use of percentiles. Percentiles help us make sense out of scores. For example, a student receiving a raw score of 57 on a statistics exam may become quite alarmed, but learning that the score fell at the 96th percentile would be of considerable comfort, especially if the instructor has promised to "curve" the students' scores.

In fact the process of "curving" grades is a dubious one, especially in course sections that are not very large. For example, grades on statistics exams do not tend to be normally distributed. Because of this and because of the frequent presence of outliers (extremely high or low values), assigning grades based on,

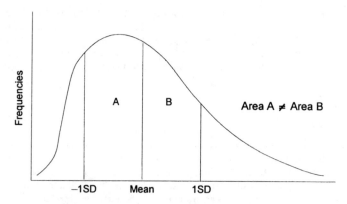

FIGURE 4.9 Comparing areas of the curve of a skewed distribution

essentially, how many standard deviations they fell from the mean can produce undesirable results.

Treating course grades as if they are normally distributed might lead an instructor to award an A to exam grades that were in the top 16 percent (above 1SD) or an F to all exam grades below the 2nd percentile (below -2SD). But is this fair? What if the class was especially knowledgeable and the 2nd percentile corresponded with a raw score of 97 out of 100? Should an individual who got a 97 on the exam get a letter grade of F, even though mastery of most of the content was demonstrated? Or should a student who got only 35 percent of questions correct get an A on an exam, just because all grades were extremely low and the 35 fell at 1SD for the class? Curving grades may be desirable if it is unknown whether an exam is too easy or too difficult. It guarantees the distribution of letter grades among class members in a way that not "too many" (whatever that is) will get any one grade, but it is unnecessary if an instructor can create a fair and rigorous exam and knows what constitutes exceptional, average, or poor performance on it.

Another common use of normal distributions can be seen in standardized tests, such as IQ tests or Scholastic Assessment Tests (SATs). Over many years, these tests repeatedly have been refined to the point that the scores of the large numbers of persons taking them tend to fall into patterns with consistent means and standard deviations. In other words, their scores form normal distributions.

SAT scores were originally designed so that combined verbal and math scores for large numbers of students would form a normal curve with a mean of 1,000 and a standard deviation of 200. In addition, all scores would fall between +3SD and -3SD from the mean. The lowest possible score would be $3 \times 200 = 600$ below the mean, or 400. (This is the 400 that one is rumored to get for just "showing up" or "signing one's name.") The highest possible (or perfect) score (the 100th percentile) would be 1,600. However, SAT scores declined considerably during the 1980s and early 1990s. Although scores of 400 and 1,600 still occur, the mean dropped to around 920. In 1994, a decision was made to adjust future test scores upward so that they would again have a mean of 1,000 and would better approximate a normal curve.

Similarly, the results of various IQ tests tend to form normal distributions. They generally have a mean of 100 and a standard deviation of either 15 or 16, depending on the test. If we understand the principles and characteristics that relate to normal distributions, it is possible, given these data, to convert any raw IQ score to its corresponding z score and then to percentile using Table 4.3. For example, a score with a z score of 1.00 (115 or 116, depending on the test) would fall at about the 84th percentile. It would also be possible to reverse this mathematical process to convert a percentile into a raw score.

Z scores are especially useful for comparing two raw scores from two different distributions when two different measuring instruments were used to measure the same variable. Z scores allow us to compare the relative position of one value (its percentile rank for its measurement) with the relative position of another value (its percentile rank for its measurement). An example follows.

TABLE 4.5 Comparative data: Two indices and clients' scores

Data	Anxiety Scale A (Gina)	Anxiety Scale B (Tom)
Raw score	78	66
Mean	70	50
Standard deviation	10	12

Example. Deborah is a social worker in a family service agency. She leads a treatment group of college students diagnosed as experiencing acute anxiety. In the past, group members have been selected for treatment on the basis of their scores on Anxiety Scale A, a standardized measuring instrument given to all students as a part of intake screening. The measuring device has a mean of 70 and a standard deviation of 10. Only students scoring over 80 on Anxiety Scale A are eligible to join the group.

A vacancy occurred in the group. Deborah checked the files of active cases in her agency and noted that the highest score among potential group members was 78 (Gina). However, Deborah had just received a referral from another family service agency stating that a new potential client (Tom) had recently moved to her city and needed further assistance. The transfer letter indicated that Tom, who suffered from anxiety, received a score of 66 on Anxiety Scale B—a different measuring instrument. The letter further stated that Anxiety Scale B has a mean of 50 and a standard deviation of 12.

Both standardized measuring devices (Anxiety Scales A and B) were valid and reliable. Based on her knowledge of normal distributions and the information received in the referral letter, Deborah saw no need to retest Tom with Anxiety Scale A. She decided to use z scores to determine whether Gina or Tom was a better candidate for the group vacancy.

To simplify her decision, Deborah constructed Table 4.5. She then computed the z score for both potential clients, which allowed her to compute the percentile for each potential group member.

$$z \text{ score (Gina)} = \frac{\text{Raw score} - \text{mean}}{\text{Standard deviation}}$$

$$= \frac{78 - 70}{10}$$

$$= .80 \text{ (corresponds to 28.81, Table 4.3)}$$

$$= 28.81 \text{ (area between raw score and mean)}$$

$$\underline{+50.00 \text{ (area left of the mean)}}$$

$$78.81$$

$$= \text{79th percentile (Scale A)}$$

$$z \text{ score (Tom)} = \frac{\text{Raw score - mean}}{\text{Standard deviation}}$$

$$= \frac{66 - 50}{12}$$

$$= 1.33 \text{ (corresponds to 40.82, Table 4.3)}$$

$$= 40.82 \text{ (area between raw score and mean)}$$

$$\underline{+50.00 \text{ (area left of the mean)}}$$

$$90.82$$

$$= \text{91st percentile (Scale B)}$$

Based on her comparative analysis using z scores, Deborah chose Tom for the group. His relatively high level of anxiety (based on his scale) made him more appropriate for the group than Gina. Furthermore, Deborah saw no need to relax the group's admittance criteria, which required a score of 80 (84th percentile), in order to admit Gina (79th percentile). Figures 4.10 and 4.11 illustrate the comparison that Deborah was able to make using z scores. Note that the score of 80 (cutoff point on Scale A) is comparable to a score of 62 on Scale B, because both fall at the point $z = 1$ (the 84th percentile). Tom's score was above this point (Figure 4.11) and Gina's (Figure 4.10) was below it.

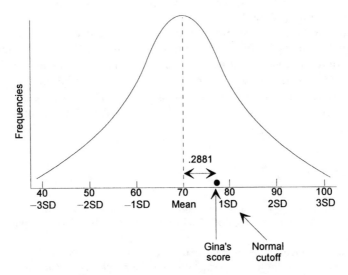

FIGURE 4.10 Distribution of scores on Anxiety Scale A (mean = 70; standard deviation = 10)

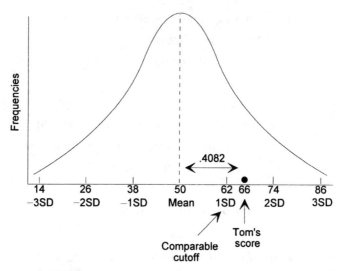

FIGURE 4.11   Distribution of scores on Anxiety
Scale B (mean = 50; standard deviation = 12)

CONCLUDING THOUGHTS

By converting raw scores from a normal distribution into z scores, we can compare individual scores in two different data sets by seeing how each score compares to others in its own group of scores. Understanding the normal distribution also enables us to visualize where a given score falls relative to others in a large sample or population.

Through the use of percentiles we can determine about what percentage of scores fall above or below any given score. This can be especially useful to the social worker in evaluating the results of standardized testing performed on clients. Even if all one knows about a test are its mean and standard deviation and the fact that it is believed to provide a valid and reliable measurement of a variable, it is possible to take a raw score on the test and put it into a meaningful perspective.

STUDY QUESTIONS

1. How does a positively skewed distribution differ in appearance from a negatively skewed one? Provide examples of variables that tend to be positively skewed or negatively skewed.
2. Discuss the characteristics of a normal, or bell-shaped, curve.
3. In a frequency polygon for the variable *number of times married* within the general population, is the distribution likely to be normal, positively skewed, or negatively skewed? Explain.

4. In a positively skewed distribution, where is the median relative to the mean?
5. With a variable that is normally distributed, approximately what percentage of all scores falls within 1 standard deviation of the mean?
6. What is the z score for a score of 79 when the mean score of all persons who complete a depression scale is 89 and the standard deviation is 5? Is this person more or less depressed than most other people who complete the scale?
7. In a normal distribution how frequently would a score occur that is more than three standard deviations above or below the mean?
8. On an IQ test with a mean of 100 and a standard deviation of 16, at approximately what percentile will an IQ of 104 fall?
9. Which z score corresponds to a higher value within a distribution of values, -1.04 or 1.00? Explain.
10. If an individual falls at the 16th percentile for weight and the 48th percentile for height, would that individual be considered underweight or overweight? Explain.
11. Discuss several ways in which a social worker can use z scores in social work practice.
12. Assume that a distribution has a mean of 12, a median of 14, and a mode of 13. Should a distribution with these central tendencies be considered normally distributed? Why or why not?
13. Use Table 4.3 to find the following:
    a. The area of the normal curve above a z score of 1.71.
    b. The area of the normal curve between the mean and a z score of -1.34.
    c. The z score that marks the lower limit of the 38 percent of the curve immediately below the mean.
    d. The z scores that mark the upper and lower limits of the middle 42 percent of the normal curve.

chapter 5

Introduction to Hypothesis Testing

The previous four chapters presented various ways to organize, display, and summarize the distribution of the values of a single variable using data drawn from a research sample or population. However, there are many times when we want to know if two variables measured within our research sample are related in some way. If they appear to be related, we may also want to know if their apparent relationship can be assumed to exist for those cases that were not in the research sample. We would like to use the findings we generated from our sample to *infer,* or make generalizations about, the population from which the sample was drawn.

What is inference? *Inference* is the degree of confidence we have when we say that an apparent relationship between two variables observed within a given research sample is a *real* one that probably exists among cases (or objects) outside the sample.

ALTERNATIVE EXPLANATIONS

In order to feel safe when making inferences, we have to be reasonably confident that the study's results cannot be attributed to other factors, called *alternative explanations,* that the research design failed to control. Three of these factors are: (a) that the apparent relationship was caused by other independent variables, (b) that the research design produced biased measurement, or (c) that the research sample was biased in some way. In research terminology, they are referred to, respectively, as (a) rival hypotheses, (b) design bias, and (c) sampling bias. Because all three alternatives are related to problems of research methodology, they must be considered before data are collected. Social work research methods texts generally devote a discussion to strategies designed to minimize their effects.

Rival Hypotheses

Rival hypotheses are really theoretical alternatives for explaining the presence of an apparent relationship between two or more variables. We must decide whether our literature review did, in fact, produce the most salient independent variables. Perhaps other variables, either individually or in concert, are the real explanation for the variations in the dependent variable. They may have caused the independent and dependent variables to appear to be related, even though they really are unrelated.

Unless the other variables that may help to explain an apparent relationship between the independent and dependent variables are measured or statistically controlled, they cannot further our understanding of their apparent relationship. Rival hypotheses are best controlled when a thorough literature review is conducted and when good judgment is used in the selection of independent variables.

Design Bias

Design bias, sometimes referred to as measurement bias, is closely related to rival hypotheses. It is a consistent distortion in the measurement of variables that can negatively affect the quality of data collected. It can produce measurements that, while perhaps reliable, are not valid.

There are many ways in which design bias can creep into a research study. For example, it might result from the fact that data were collected at an atypical time, or that the research study was influenced by some outside event, or to the conscious or unconscious tendency of the person collecting the data to get a less-than-true perception of what was occurring. If, for whatever reason(s), the measurement of variables is biased, two variables can appear to be related when, in fact, they are not.

Design bias is best eliminated or at least minimized by using a good research design that incorporates sound measurement practices. A true experimental design provides good control for this alternative explanation through randomization, but true experimental designs are rare in social work research.

Even with other types of research designs, it is sometimes possible to use statistical analyses to detect the presence of measurement bias within data that have been collected. However, at that point, it may be too late to eliminate its negative effects on the quality of a research study.

Sampling Bias

Sampling bias is the systematic distortion of a sample. It can occur either intentionally or unintentionally. It can result from factors such as the sampling method(s) that was used to select it, to when, and where it was selected. A biased sample can be characterized by overrepresentation (or underrepresentation) of some types of cases (relative to the population from which it was drawn).

How can we be reasonably certain that sampling bias is not the real explanation for an apparent relationship between two variables? By using a method to select a sample that will increase the likelihood that its characteristics are the same as those of the larger population from which it was drawn. We can pay special attention to how, when, and where it is selected and can use sampling strategies (e.g., randomization, matching, stratification) that are designed to increase the sample's representation from the population from which it was drawn.

CHANCE

Even if all three previously discussed alternative explanations for an apparent relationship between or among variables within a sample can be dismissed, there is another possiblity—*chance* (often referred to as sampling error). It has nothing to do with a study's research design; it contends simply that the apparent relationship was just a "fluke." It exists among only those cases within the sample. Even the best research study, using the best methodology, the best operationalization of variables, and the best data analysis techniques, can never totally rule out chance as the explanation for an apparent relationship between or among variables.

In order to conclude that an apparent relationship between the variables within our sample data reflects a real relationship within a population, rival hypotheses, design bias, and sampling bias must be convincingly dismissed as alternative explanations. Even if we can rule out these alternative explanations, we are still left with the possibility that the relationship occurred just by chance. Chance is usually the final explanation that the skeptical consumer of a research report proposes as the real reason why two variables appear to be related.

How do we determine what role chance may have played in an apparent relationship between or among variables? Unlike rival hypotheses, design bias, and sampling bias, solid research designs are not sufficient to minimize the effects of chance. There are other ways to address the possibility of its presence, which involve statistical analysis. However, before we talk about the role of chance, we need to discuss briefly probability theory.

Probability Theory

Probability theory relates to the mathematical likelihood of an event occurring. It is based on certain rules, sometimes called laws. For example, one rule states that the likelihood of any one event occurring can range from 0 (never) to 1.0 (absolute certainty).

Central to probability theory is the assumption that while certain patterns of events can be seen to exist in many repeated observations over time, individual, or short-term, observations tend to differ somewhat from the overall long-term patterns. For example, if we flip a fair coin in the air, it has a 50-50, or 50 percent, chance of landing on heads. Obviously, it also has a 50 percent probability of landing on tails. However, this is a *theoretical* probability. It suggests the percent of heads (or tails) that would occur in an infinite number of flips of the coin, when the "law of averages" has had an opportunity to take effect.

In reality, we know that if we flip a coin ten times, we can get a different result from the one that is expected to occur—that is, five heads. We would not be surprised if we obtained four, six, or even three or seven heads. We would simply blame it on chance and assume that, in the long run, if we repeated the coin tossing many more times, the percent of heads would eventually approximate 50 percent.

What if the ten coin flips produced nine or ten heads? At this point, we might suspect that something was amiss. The results seem to be so unlikely that they "defy the laws of probability." How could these results have occurred? Could the coin be defective? Could the way in which the coin was tossed have influenced the results? What is going on here? Another way of stating these questions might be, "Is something else (some other variable) influencing the results (the number of heads)?"

The laws of probability have been applied in many areas of life. For example, the *addition rule* of probability would tell us that, since the probability of a student guessing correctly on a multiple choice question with four alternative answers is $1/4$, or .25, the likelihood of getting one answer correct while guessing on two questions is $1/4 + 1/4 = 1/2$, or .50. Another rule, the *multiplication rule,* would tell us that the likelihood of guessing correctly on both questions is $1/4 \times 1/4 = 1/16$, or .625. Mathematicians use the various laws of probability to determine what *should* happen in various situations, that is, what would happen in the long run, based upon chance. Of course, we know that, in the short run, strange aberrations can and do occur. For example, on a single examination a student might guess on two questions and get both or neither one correct.

Probability and the Normal Distribution

Another way to understand probability is by way of the normal distribution. In Chapter 4 we discussed how to compute a raw score's percentile using z scores. The percentile score can also be interpreted as a probability score where the total area under the normal curve (from minus infinity to plus infinity) represents a probability of 1.0. Suppose we collected age data on adolescent clients participating in an anger management program. The 14 adolescents in the program consist of two 12-year-olds, three 13-year-olds, four 14-year-olds, three 15-year-olds, and two 16-year-olds. The ages of these clients is normally distributed; the mean, mode, and median age are all 14.

If we were to draw a sample of one case from the group of adolescents, there would be a 100 percent chance, or a probability of 1.0, of selecting a client who is between the ages of 12 and 16. That is because all adolescent clients are represented in the distribution. We can reduce this probability by specifying a particular range. What if we wanted to know the probability of randomly selecting a 13- or a 14-year-old client from our sample distribution? The probability would be 3 chances out of 14 for picking a 13-year-old and 4 chances out of 14 for choosing a 14-year-old—a total of 7 chances out of 14 (.50) for randomly selecting a 13- or a 14-year-old. Thus we would have a 50 percent chance of selecting a 13- or 14-year-old from our sample. Like units of standard deviation, intervals, or cutoffs scores, correspond to areas of space under the normal curve. Narrow intervals correspond with smaller areas under the normal curve and result in a lower probability of having a value fall within the specified range. Wider intervals correspond with larger areas under the normal curve and result in a higher probability of having a value fall within the specified range.

REFUTING CHANCE

There are basically two different ways to demonstrate that an apparent relationship between two variables within sample data was not the result of chance. They are replication and statistical analysis.

Replication

Suppose we observe within sample data taken from 20 medical social workers that female social workers had higher levels of job satisfaction than male social workers. In an ideal world with unlimited research resources, we could attempt to determine if the apparent relationship within the sample data was a "fluke" through replication. *Replication* refers to repeating the research study using the same methods to see if the same findings result. We could simply draw another sample of medical social workers and see if the relationship between gender and job satisfaction occurred again. Perhaps we could draw 100 different samples and see how many times out of 100 the same relationship occurred. Then if, for example, the relationship occurred at least 95 times out of 100, we might conclude that the relationship is a real one.

Of course, in the real world, repeated replication usually is impractical. Obtaining resources and access to data even once is often difficult for the social work researcher, so how could we ever be able to do it 100 times? Some replication occurs in social work research, however, it is used primarily to attempt to repeat the results of research studies where a relationship between variables was already found to exist. It is used far less to demonstrate initial support for the existence of a relationship.

While the cost of repeated replication makes it very impractical, other methods for determining if an apparent relationship exists between variables employ some of the principles of replication. Later in this chapter we shall see how the notion of "95 times out of 100" is used in making decisions based upon the statistical analysis of the data.

Statistical Analyses

Let us return to our example. Another way that we could gain logical support for the inference that there was a relationship between gender and job satisfaction is if we had made an educated guess (prediction) that we would find one.

Suppose that we had arrived at this prediction through the process of synthesizing existing qualitative and quantitative knowledge on gender differences and job satisfaction among medical social workers. We used our own observations and experiences in working with medical social workers, as well as the writings of scholars in the profession and many other sources, such as unpublished documents and interviews with persons who "ought to know." In other words, we had conducted an extensive literature review and synthesized the knowledge thus found with our own observations. Based upon what we learned, we predicted that female medical social workers would have higher levels of job satisfaction than male medical social workers.

Suppose that the two variables were found to be related within our sample, exactly as we anticipated they would be. Would that not convince us that the relationship we observed was not just the work of chance? It would help, but we would need more convincing evidence. Even if we could rule out the alternative explanations for an apparent relationship (rival hypotheses, design bias, and sampling bias), and even if we had predicted our results based upon existing knowledge, it still might have been the work of chance. We would have to more rigorously "test" our belief and thus demonstrate that the results are so impressive—so clearly in support of it—that the relationship between the variables must be a real one within the population from which the sample was drawn. We can do this by stating and testing hypotheses using statistical analyses.

RESEARCH HYPOTHESES

As discussed briefly in Chapter 1, our belief about the relationship between two variables can be stated as a hypothesis. There are many different definitions of hypotheses, but they all suggest the same idea: A hypothesis can be thought of as a tentative answer to a research question. It also is a statement of a relationship between or among variables.

Hypotheses are generated in different ways. They may evolve as the product of someone else's research study, which employed inductive methods. This occurs most frequently in exploratory or formulative research studies in which more "qualitative" methods are used. Hypotheses may also evolve in a deductive way from our own observations and our review of the literature. Among other things, we use the literature to narrow or refine a general research question. Frequently, as the evidence in the literature starts to accumulate, we think we may have an answer to that question. We then try to express our tentative impressions or our conclusions in the form of a hypothesis.

A *research hypothesis,* also called a substantive hypothesis or an alternative hypothesis, is a statement that expresses what we believe to be the real relationship between or among certain variables. It is a hypothesis that we wish to test, by using a suitable research design developed with the help of a review of literature. There are three basic types of research hypotheses: (a) one-tailed hypotheses, (b) two-tailed hypotheses, and (c) null hypotheses.

One-Tailed Hypotheses

A *one-tailed hypothesis* states that there is a specific relationship between or among variables. It predicts which values of one variable will be found to be associated with which values of the other variable(s). Continuing with our previous example, a one-tailed hypothesis might be, "Male social workers have higher levels of job satisfaction than female social workers" or "Female social workers have higher levels of job satisfaction than male social workers." It is important to note that a one-tailed hypothesis predicts the direction of a relationship between two variables, which is why it is sometimes called a *directional hypothesis.*

Two-Tailed Hypotheses

A *two-tailed hypothesis,* also referred to as a *nondirectional hypothesis,* states only that there is a relationship between or among variables. Unlike the one-tailed hypothesis, it does not predict which values of one variable will be associated with which values of the other variable. A two-tailed hypothesis for our example might be, "Gender is related to job satisfaction levels." Note that it does not predict whether males or females will be found to have higher levels of job satisfaction, as the one-tailed hypothesis did.

Null Hypotheses

A *null hypothesis* states that there is no relationship between two variables. The null hypothesis for our example would be, "There is no relationship between gender and job satisfaction." A null research hypothesis is, in a sense, the opposite of either a one-tailed or a two-tailed hypothesis.

Research hypotheses stated in the null form are relatively rare. They are used when we wish to gain support for the belief that two variables believed by some people to be related really are unrelated. Null research hypotheses are used to try to dispel false beliefs. For example, in the past, researchers sought to disprove the racist stereotype that one race is intellectually superior to another. They did so by finding statistical support for the null hypothesis (race and intelligence are *not* related). Similarly, researchers sought to demonstrate that women are just as effective fire fighters as men. They set out to find support for the null research hypothesis that there is no relationship between gender and fire-fighting competence.

Even though it is somewhat unusual to see a research hypothesis stated in the null form, the *idea* behind the null hypothesis is important to all hypothesis testing. The null hypothesis plays the skeptic. It suggests that, even though two variables may appear to be related within a set of data, they really are unrelated— there are other explanations for their apparent relationship.

How is the idea behind the null hypothesis helpful if we are trying to find support that variables really *are* related? To gain support for a hypothesis that two variables are related (using a one-tailed or a two-tailed hypothesis), we must first demonstrate that they are *not unrelated.* To put it another way, we must demonstrate that it is safe to conclude that the null hypothesis is wrong—that is, that the variables really are related.

When we feel confident that two variables really are related, we say that we can "reject the null hypothesis." To be able to reject the null hypothesis, we need to demonstrate that chance is an unlikely explanation for the apparent relationship between or among variables and that a true relationship is a more plausible explanation.

A one-tailed or two-tailed research hypothesis is always tested indirectly—that is, its null hypothesis is what is statistically tested. In short, statistical tests are used to help determine whether or not it is safe to reject the null hypotheses.

In research, we do not refer to research hypotheses as proven or disproven; they are either declared as "supported" or "not supported," a conclusion based upon whether or not we feel it is safe to reject the null hypothesis. In our professional literature, we often see it reported that a research hypothesis was supported through some form of data analysis. This really means that a null hypothesis about the relationship between or among certain variables was rejected.

TESTING RESEARCH HYPOTHESES

Once data have been collected, we are in a position to organize and summarize them using the procedures presented thus far in this book. As we do this, we may begin to sense whether or not our research hypothesis is supported by the data we have collected. There may or may not be an apparent relationship between or among variables within the sample data that were collected.

Let us return to our previous example to illustrate how this process might work. Suppose that, after a thorough literature review, we decided that we had justification for stating the following one-tailed research hypothesis:

> Female medical social workers will have higher levels of job satisfaction than male medical social workers.

Remember, it is the null hypothesis that is statistically tested—not the research hypothesis. The null hypothesis for the above one-tailed hypothesis would be:

> There is no relationship between the gender of medical social workers and their levels of job satisfaction.

We then randomly selected all 20 medical social workers employed at a local hospital. We noted each social worker's gender and administered to each one the same standardized measuring instrument designed to measure job satisfaction—a job satisfaction scale. The instrument is standardized so that scores can range from a maximum score of 100 (high job satisfaction) to a minimum score of zero (low job satisfaction).

As we analyzed our data, frequencies might seem to provide a reason to reject (or not to reject) the null hypothesis, particularly since the number of social workers was small and it would be easy to identify any patterns in the distribution of the dependent variable (job satisfaction) between the two genders. If, for example, we observed that the two groups (females and males) *both* had an average job satisfaction score of 60, we would obviously not reject the null hypothesis (there is no difference between the mean satisfaction scores for males and females). However, if the ten females were found to have an average job satisfaction score of 80 and the ten males had an average score of 20, we would feel that we had pretty strong support for the rejection of the null hypothesis.

What if there was a difference in job satisfaction between the two subsamples (males and females), but it was not particularly dramatic? What if the mean job satisfaction was 62 for females and 56 for males? How likely is it that the observed relationship is a real one that represents a relationship between the two variables within the larger population (all medical social workers) from which the sample was drawn? Could we infer, based on these central tendency data alone, that, as a group, female medical social workers not in the research sample possess a higher level of job satisfaction than males? Such a conclusion would seem a little premature, even if we had carefully designed and implemented our research study so that we were fairly certain that rival hypotheses, design bias, and sampling bias were ruled out as alternative explanations for the apparent relationship. However, what about chance?

Statistical analyses attempt to discredit chance as an explanation for an apparent relationship between two variables within sample data. *Inferential tests* (such as those that we will examine in later chapters) tell us how safe we would be if we were to infer that an apparent relationship between two variables within sample data is a real one that exists beyond the research sample. While they differ from each other in their approach to this task, inferential statistical tests also have many similarities. They all determine the likelihood that an apparent relationship between two variables has occurred as a result of chance. If it is highly unlikely that chance lies behind the apparent relationship, and if rival hypotheses, design bias, and sampling bias have been eliminated as alternative explanations, only one reasonable explanation is left—a true relationship is believed to exist in the real world. It probably would be safe to reject the null hypothesis.

In attempting to gain support for a research hypothesis, we can never totally eliminate chance as the explanation for an apparent relationship between or among variables. However, we want to be reasonably certain that what we observe is not a fluke occurrence. We do not want to report a relationship between or among variables that appears to be real when it is not. At the same time, we do not want to be so rigid or unreasonable that we will not claim support for a relationship just because there is a remote possibility that chance may have produced it. If we did that, few if any research findings would ever see the light of day.

ERRORS IN DRAWING CONCLUSIONS
ABOUT RELATIONSHIPS

Ultimately, we must decide whether to reject the null hypothesis or not. We must do this based on available data, the methods used to acquire them, and the statistical analysis that was performed upon them. Two types of errors, Type I and Type II, can be made in interpreting research data. A *Type I* error occurs when we reject the null hypothesis and conclude that a relationship between the two variables really exists when, in fact, no true relationship exists. A *Type II* error is just the opposite. It occurs when we fail to reject the null hypothesis and fail to identify a true relationship between two variables when one exists. The two types of error are illustrated in Table 5.1.

Type I and Type II errors can result from many factors relating to the research design of a study. These include selecting a biased sample, utilizing data collection instruments that are invalid and/or not reliable (design bias), and failing to control for the presence and effect of other variables (rival hypotheses). Anything in a research design that can produce flawed data and, thus, misleading conclusions drawn from the data can cause Type I and Type II errors.

Type I and Type II errors can also result from the use of inappropriate statistical tests. If, for example, we incorrectly use a statistical test that requires certain conditions that are not present, or we employ a test that is too "weak" to detect a real relationship between two variables, Type I or Type II errors can occur. In the first instance, the data were treated as if they possessed qualities they really lacked; in the second case, the opportunity to use a more exacting analysis was not exercised. If the appropriate statistical test is not used, we can falsely conclude that a true relationship between two variables exists (Type I error) or a true relationship may remain unidentified (Type II error).

We can never totally eliminate the possibility of committing an error when we make the decision either to reject the null hypothesis or not to reject it. In fact, if we are overly careful not to commit a Type I error (mistakenly rejecting the null hypothesis), we increase the likelihood of committing a Type II error (mistakenly failing to reject the null hypothesis). Conversely, if we are overly careful to avoid committing a Type II error, we increase the likelihood of committing a Type I error. We must ultimately decide which error, Type I or Type II, is more acceptable to us, should it occur. This is, in part, an ethical decision that

TABLE 5.1 Type I and Type II errors

Real World	Our Decision	
	Reject Null Hypothesis	Do Not Reject Null Hypothesis
Null hypothesis false	No error	Type II error
Null hypothesis true	Type I error	No error

requires a knowledge of social work practice and the consequences of committing one error over the other. Fortunately, as we shall see, there are also statistical conventions to guide us in this decision-making process.

In some research studies, the consequences of Type I or Type II errors can be potentially grave. For example, if social work practitioners do not recognize the misuse of a statistical test or some methodological error has resulted in a Type I error, they may mistakenly conclude that there is a true relationship between a particular treatment method and an increased rate of client success. They may adjust their treatment approaches based on this result. They may also respond to other research findings in which, for some reason, a Type II error was committed and discard a treatment method that was really effective but appeared not to be related to client success.

A Type I error is not inherently more or less desirable than a Type II error. When we apply research findings to social work practice situations, both Type I and Type II errors have the potential to harm our clients or to result in a wasteful expenditure of limited agency resources.

Even if research studies are well designed and we understand and apply the criteria for appropriate selection of a statistical test, there is always some possibility, no matter how "acceptably" small, of committing an error in drawing conclusions from the data that we have collected. It always remains possible that we have happened upon that one-time-in-a-million sample that leads us to draw an erroneous conclusion about the relationship between or among variables within the population from which the sample was drawn. We may also have made some obscure methodological error that failed to control for the effects of other variables, introduced design bias into our research design, or resulted in a biased sample. This remote possibility should not, however, preclude us from taking reasonable risks in interpreting research findings and drawing conclusions and implications from them. This is how we make progress in becoming knowledge-based social work practitioners.

STATISTICAL SIGNIFICANCE

As we noted earlier in this chapter, it is usually impractical to repeat a research study (that is, replicate it) enough times to convince us whether or not there is a real relationship between or among variables. It would be nice, for example, to repeat the research study in our earlier example using 100 different samples of medical social workers in 100 different hospitals around the country to see if females showed higher levels of job satisfaction than males in most of the samples (say, in at least 95 percent of the samples). Typically, we get only one shot to do a research study and collect the necessary data. Thus, within a single research study, we need some comparable evidence that a relationship between or among variables is a real one. At what point can we be sufficiently certain that whatever apparent relationship we find in our sample cannot reasonably be dismissed as the work of chance? We must rely on mathematics, common sense, and convention.

Rejection Levels

As we noted earlier, the decision to reject the null hypothesis does not totally rule out chance as the explanation of an apparent relationship between two variables. It takes the position that some methodological problem or chance may have caused the variables to appear to be related when, in fact, they were not, but the likelihood of this is sufficiently remote. Rejecting or not rejecting the null hypothesis involves risk (of a Type I or Type II error, respectively), but the risk is supposed to be a small one. We should be fairly comfortable with the decision, confident that it is the correct one.

Over the years, most researchers have settled on the 95 percent certainty level as the point at which they are sufficiently confident to be able to reject the null hypothesis. Expressed another way, they feel safe in concluding that two variables are related if a statistical analysis suggests that there is less than a 5 percent chance that, if the null hypothesis were to be rejected, a mistake would be made. That much risk of committing a Type I error is acceptable in most research studies. This is referred to as the .05 rejection level. *Rejection levels* are also sometimes called alpha levels, probability levels, and significance levels.

There is nothing sacred about the .05 rejection level. While it is the most widely used for rejecting the null hypothesis, other rejection levels also can be used. The decision to use levels other than .05 is based on our assessment of the possible consequences of mistakenly rejecting the null hypothesis or failing to reject it. A more demanding proof of a relationship between two variables, such as a .025 or .01 rejection level, might be used when we wish to allow only a very small possibility that we would reject the null hypothesis in error and conclude that a relationship exists between two variables when chance is the real explanation for the apparent relationship between them (a Type I error). These rejection levels allow for even less likelihood that chance is the reason for an apparent relationship than does the conventional .05 level. If the application of the research findings might be a matter of life and death, such as in research studies on a new drug to treat HIV patients, an even more demanding threshold for rejection of the null hypothesis might be used, such as .001. The .001 rejection level means that the probability of erroneously rejecting the null hypothesis is less than 1 out of 1,000.

In research studies in which the consequences of mistakenly rejecting the null hypothesis are less likely to be fatal or traumatic, we occasionally consider a .10 rejection level as acceptable. A .10 rejection level allows for twice the possibility of committing a Type I error because of chance (in the form of sampling error) than does a .05 level. It also reduces the likelihood of committing a Type II error. Sometimes a less demanding rejection level, such as a .10, is used as evidence of a relationship between or among variables if the research design includes at least one replication. While achieving one .10 rejection level may be viewed as inconclusive support for a relationship, achieving it two or more times in succession may lead us to the conclusion that the null hypothesis can be rejected.

While some flexibility is allowed in selecting the threshold at which chance is reasonably eliminated as the explanation for an apparent relationship, the choice of a rejection level should not be viewed as casual. Convention states that the .05 rejection level should be used unless we develop and state a convincing rationale for the use of another level. The selection of a rejection level must also be made *before* data are collected. It would be unethical to change the level afterward, because the decision could be construed as an effort to manipulate the findings to support the researcher's beliefs.

When reporting the findings of a statistical analysis, the rejection level that was used as a "cutoff" to reject the null hypothesis should be reported. Usually, this is done using wording like, "The average job satisfaction level for females was 85 and for males 65. This 20-point difference is statistically significant ($p <$ .05)." The p is the rejection level and is always in lower-case italics. The p stands for probability and should always be reported in a research report. The value of p is usually presented by a "less than" sign ($<$); however, if statistical significance is not achieved, it is reported using the "greater than" sign ($>$). What we would be reporting in the above statement is that the probability that the relationship could have occurred by chance is less than 5 percent.

Locating Rejection Regions in Normal Distributions

Many statistical tests compare data drawn from a sample with the standard normal sampling distribution table (i.e., Table 4.3 or Appendix A). They test hypotheses by attempting to determine whether the distribution of a variable within a sample is or is not typical of what we might expect to get from simply drawing a sample from a population. We can use the concept of rejection regions to determine just how typical or atypical our sample is in relation to the variable. The term *rejection region* refers to the region(s) of a normal curve that suggest it is safe to reject the null hypothesis. In order to understand the concept of rejection regions, we must rely heavily on the concepts of normal distributions and z scores that were developed in Chapter 4.

Rejection Regions for Two-Tailed Hypotheses. The rejection region for statistical tests that test hypotheses in the way that we have just described depends on (a) the rejection level selected and (b) whether the research hypothesis is two or one tailed. For two-tailed hypotheses, the z values of the two rejection regions are located using procedures similar to those discussed in Chapter 4 that we used to locate a raw score's standardized z score. The steps for locating the two rejection regions for two-tailed hypotheses are listed below:

1. Divide the selected rejection level by 2.
2. Subtract the derived value from .5000 and multiply by 100.

3. Locate the derived proportion in the body of the normal distribution table (i.e., Table 4.3 or Appendix A).
4. Determine the z score that corresponds to that proportion.

If we are using the conventional .05 rejection level, the calculations for the preceding four steps would be:

1. $^{.05}/_2 = .025$.
2. $(.5000 - .025) \times 100 = 47.50$.
3. Find 47.50 in the body of the normal distribution table (i.e., Table 4.3 or Appendix A).
4. The z score corresponding to 47.50 is + or −1.96.

How do we interpret the results of the above calculations? They mean that when we are using a rejection level of .05, the rejection region for a two-tailed research hypothesis is the area greater than or equal to a positive (+) or negative (−) z score of 1.96. Such a result must be achieved in order to reject the null hypothesis. Figure 5.1 provides a graphical illustration of this concept.

It should be noted that + or −1.96 is the required z value for *all* two-tailed hypotheses that use the .05 rejection level. So long as the standard normal distribution table is used (i.e., Table 4.3 or Appendix A), it will always be the same.

Rejection Regions for One-Tailed Hypotheses.   Since one-tailed hypotheses predict the direction of a relationship between or among variables, there is only one variation in the above procedure for obtaining their appropriate rejection region—the selected rejection level is not first divided by 2. The .05 rejection region for a one-tailed research hypothesis is derived as follows:

1. $(.5000 - .05) \times 100$.
2. Find 45.00 in the body of the normal distribution table (i.e., Table 4.3 or Appendix A).
3. The z score corresponding to 45.00 is + or −1.65.

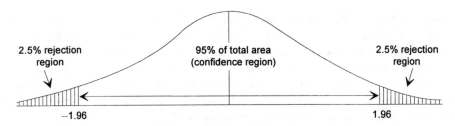

FIGURE 5.1   The normal distribution showing the 95% confidence region and two 2.5% rejection regions for two-tailed research hypotheses

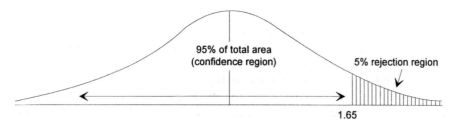

FIGURE 5.2 The normal distribution showing the 95% confidence region and the one 5% rejection region for an upper one-tailed research hypothesis

How do we interpret the results of the previous calculations? When using a one-tailed hypothesis and a rejection level of .05, the calculated z must be greater than or equal to a positive (+) or negative (-) z score of 1.65 in order to reject the null hypothesis. Either + or -1.65 (depending on the direction of the hypothesis) is the required z value for all one-tailed hypotheses that use the conventional .05 rejection level and the standard normal distribution table (i.e., Table 4.3 or Appendix A).

Figure 5.2 graphically portrays the rejection region (z = at least 1.65) for a one-tailed research hypothesis where the upper tail is specified, that is, where the predicted direction of relationship between variables is positive. Figure 5.3 provides an illustration of the rejection region (z = at least -1.65) for a one-tailed research hypothesis where the lower tail is specified, that is, where the predicted direction of relationship between variables is negative. Unlike Figure 5.1, where there is no direction predicted and, thus, there are two rejection regions, Figures 5.2 and 5.3 contain only one rejection region, the one that reflects the direction of the hypothesis. Only z scores that fall within that region will allow us to reject the null hypothesis.

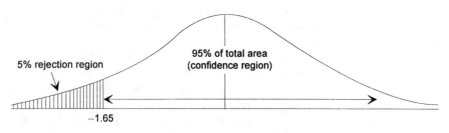

FIGURE 5.3 The normal distribution showing the 95% confidence region and the one 5% rejection region for a lower one-tailed research hypothesis

By glancing at Figures 5.1, 5.2, and 5.3, it can be observed that a one-tailed research hypothesis has certain advantages. It moves the critical rejection region closer to the mean of the sampling distribution, thus improving the probability of rejecting the null hypothesis. This is only logical, if we think about it. When we predict the direction of a relationship using a one-tailed hypothesis, we should get a little "extra credit" if support for the relationship that we predicted is found. As we noted earlier, a specific relationship that has been predicted based upon existing knowledge is less likely to be the work of chance than one that we did not specify in advance. Note, however, that if the wrong direction has been specified in a one-tailed hypothesis—that is, if we incorrectly predicted the direction of a relationship using a one-tailed hypothesis—the probability of the rejection of the null hypothesis is zero. We are given no prediction "credit" (and we should not be) for results that were the exact opposite of what we predicted. Thus, there is some risk involved in using a one-tailed research hypothesis. It should not be used *unless:* (a) the direction of a relationhip can be confidently predicted based upon available knowledge, or (b) the researcher is interested only in those sample outcomes that fall in one tail (direction) of a sampling distribution of a variable.

STATISTICALLY SIGNIFICANT VERSUS MEANINGFUL FINDINGS

The word *significant* is widely and loosely used in our profession and elsewhere to emphasize the importance of something, such as when we refer to a social worker's "significant contribution" to the passage of some social legislation or the role of a "significant other" in the development of a client's self-esteem. As with several other words that we use daily, it is best to set aside the everyday meaning of *significance* in order to understand its specific meaning in the field of statistical analysis.

Statistical significance is the demonstration, through the use of statistical testing, that an apparent relationship between or among variables is unlikely to have been produced by chance. A relationship that is declared to be statistically significant is one that we are reasonably certain—95 percent, in most instances—cannot be explained away as a fluke. For us, this is the only relevant meaning of the word *significant* or *significance;* we must be careful to use the terms in this sense, and only in this sense.

The presence of a statistically significant relationship between or among variables, of course, still may be dismissed as just the result of rival hypotheses, design bias, or sampling bias. A statistical analysis, as we have suggested earlier, is useful only if the data that it is calculated from are valid and reliable. However, what if these three alternative explanations (for the existence of a real relationship between variables) can be successfully dismissed? In that case, does the presence of a statistically significant relationship demonstrating that chance is a very unlikely explanation for the apparent relationship mean that the relationship probably is a real one?

Yes! The relationship probably *is* real; the variables probably really *are* related. But is the relationship between them necessarily a meaningful one? Maybe, maybe not. We must be careful to evaluate every statistically significant relationship between two variables in the context of the question, "So what?" In social work practice, every statistically significant relationship is not a meaningful finding that cries out for implementation. In fact, some statistically significant relationships are meaningless in the absolute sense.

Example

An example may help to illustrate this distinction. Suppose that Angie, an administrator of a family service agency, conducts a research study using a large sample of clients to determine which type of treatment approach (A or B) produces better marital adjustment when used with couples in marital counseling. She lacks clear indication as to which is more effective, so she uses a two-tailed research hypothesis: "There will be a statistically significant difference in marital adjustment between clients in Treatment A and those in Treatment B."

Angie believes that her measurement of all variables (including marital adjustment) is valid and that her research design has adequately controlled for the effects of all three alternative explanations of her findings: rival hypotheses, design bias, and sampling bias. Using a standardized marital adjustment scale, she finds an average marital adjustment level of 51 among couples who received Treatment A and an average level of 54 among couples who received Treatment B. (Higher scores indicate better adjustment.)

Using an appropriate form of statistical analysis, Angie concludes that the difference is statistically significant. The relationship between type of treatment and level of marital adjustment is apparently a real one, and it allows her to reject the null hypothesis. However, despite the finding of statistical support for her research hypothesis, Angie concludes that the findings are not very meaningful, given her current budget constraints. An average difference of three points (54 − 51 = 3) on the marital adjustment scale is really quite small in the absolute sense. It is not large enough to justify sending her staff members to an expensive staff development workshop to get the knowledge and skills necessary to use Treatment B. Based on the lack of what can be viewed as a meaningful difference in the results of the two types of treatment (at least for Angie in her current situation), she decides that the findings, while interesting and statistically significant, should not be implemented at the present time.

The existence of a statistically significant relationship between two variables can be determined by statistical testing, which is based on the laws of mathematical probability. However, the determination of whether a finding is meaningful or not requires human judgment; it entails the use of insight into many different aspects of the social work practice milieu.

CONCLUDING THOUGHTS

Research designs are the primary means to eliminate three possible explanations of apparent relationships between variables that we may observe in data collected from a research sample (rival hypotheses, design bias, and sampling bias). Statistical testing is used to tell us the probability that an apparent relationship may have been produced by chance. In deciding whether or not to reject the null hypothesis or, if we do reject it, whether we have a finding that is meaningful, we must employ common sense. These decisions must always be made with reference to their potential to benefit or to harm social work clients.

STUDY QUESTIONS

1. Before we can claim a true relationship between or among variables, what competing explanations for an apparent relationship must be eliminated?
2. Which one of the competing explanations do statistical procedures seek to discredit?
3. What competing explanations are controlled primarily by the design of a research study?
4. What are some other terms for chance that are sometimes used?
5. What is the difference between a Type I and a Type II error? How does reducing the likelihood of committing one affect the likelihood of committing the other?
6. What is the null form of a statement of a relationship between the variables *age* and *political party preference?*
7. What is the relationship between the null hypothesis and chance in hypothesis testing?
8. Does a "statistically significant" relationship between two variables mean that there is no possibility that the variables are unrelated? Explain.
9. When might we use a rejection level other than the conventional .05 to conclude whether statistical support exists for a research hypothesis? Provide examples from potential social work research studies.
10. Which significance level, .01 or .10, suggests a greater likelihood of a true relationship between two variables? Explain.
11. Discuss the advantages and disadvantages of implementing the results of a study that found a 10 percent average difference in job satisfaction between the female and male social workers in a hospital setting. Discuss what additional information you would want before taking action on the study's results. Justify any assumptions or recommendations.
12. Discuss what you would do with a research study's finding that there was a significant difference ($p < .05$) between the average client hospital readmission rate for those workers who used Treatment A and those who used Treatment B. How would you use the fact that Treatment A had an average client readmission rate of 15 percent and Treatment B had an average client readmission rate of 17 percent? Or the fact that Treatment B requires 25 percent more staff than Treatment A? Justify your answer. Does statistics help us to make an ethical decision in this situation? Why or why not?
13. As an administrator of your local United Way agency, your job is to allocate funds for agencies that request them. Two agencies (A and B) require money from you so that they can stay in operation next year. Both offer the same services, but you have enough money to fund only one. Agency A states that its clients had an average

treatment success score of 42 on a standardized measuring instrument that measures client functioning ($N = 1,000$). Agency B, using the same measuring instrument, states that its average treatment success score was 44 for the same period. Agency B also had 1,000 clients. You take these data and calculate the appropriate statistic to determine if chance played any role in the difference in scores. Your statistical analysis produces a finding of $p < .25$. In other words, a two-point average difference on the social functioning scale would happen less than 25 times out of 100, or less than one in four, with two groups of 1,000 clients each. Will you be able to use your statistical analysis to decide which agency to fund? Why or why not? Discuss what additional information you would want before taking action on the study's results. Justify your answer.

14. You are an administrator in a very large family service agency. There are two treatment techniques (A and B) that your workers use to help parents whose children have school truancy problems. Half your workers use Treatment A and the other half Treatment B. You conduct a research study to determine which treatment is more effective in reducing truancy. When they first applied for services, all children in the study were truant an average of five times per month. After treatment, the parents who received Treatment A reported an average of two instances of truancy per month, while the parents who received Treatment B reported an average of three instances of truancy per month. The difference of one is statistically significant at the .05 rejection level. How could you use these findings?

chapter 6

Selecting Statistical Tests

Chapter 5 presented a theoretical discussion of how statistical testing is used to test research hypotheses. It described how statistics are used to determine the probability that chance may have caused variables within a hypothesis to appear to be related when they really are not. This chapter will complete the discussion on hypothesis testing by presenting a few of the basic issues that need to be addressed when selecting the best statistical test to test a research hypothesis.

THE IMPORTANCE OF SELECTING
THE CORRECT TEST

The many decisions that need to be made during the course of a research study have the potential to enhance—or to harm—the credibility of the study's findings. For example, a biased literature review, the selection of an invalid measuring instrument, the use of a biased data collection method, or the use of an ineffective sampling method can all cause readers to doubt the findings derived from a research study. Even if the study was well designed and implemented, the credibility of its findings still can be jeopardized by one more error—the selection and use of the wrong statistical test. A critical reader of a research report always asks, "Was an appropriate statistical analysis used to test the research hypothesis?" If not, even the best executed study is likely to produce only skepticism on the part of the reader. Unfortunately, it is not unusual for research studies to suffer from this fate.

Why are inappropriate statistical tests sometimes chosen when appropriate ones are available for almost any type of data analysis? One reason may be the "rule of the instrument," which states that some people have a tendency to see the solution to any problem as requiring what they know and/or do best—that is, that with which they are most comfortable and familiar. For example, marriage counselors may see the solution to a problem such as child abuse as more marriage counseling services, whereas family therapists may see a need for more family treatment. On the other hand, lawyers may see improved legal services as the solution, whereas politicians might think the problem could best be addressed through legislation.

Many social workers, particularly those who received their formal social work education 20 or more years ago, are unlikely to have extensive background in statistical analyses. Their knowledge often is confined to a passing familiarity with only one or two statistical tests. Facing the necessity of choosing a statistical test to examine their research hypotheses, they are likely to fall prey to the "rule of the instrument." They turn to an "old friend" with which they are most familiar rather than exploring the possibility of using a newer or more appropriate test that would require additional study.

There also is a widely held misconception that, because statistical tests have so much in common, it makes little difference which one is used. After all, are they not just slight variations of one another? Why not just use a test that is widely used and understood by both researchers and the potential consumers of the study's findings? Why bother to seek out one that may be more appropriate but is less widely known?

There are two very good responses to these questions. First, as suggested earlier, using an inappropriate statistical test can result in a loss of credibility for the study's findings among consumers of a research report who are knowledgeable about statistics. Second, and most importantly, consumers of a research report who are not knowledgeable about statistics may fail to question a study's findings and recommendations that *should* be challenged.

There are hundreds of statistical tests available today. How do we determine which one(s) to use? This decision is relatively complicated, and several factors must be considered. The remainder of this chapter will examine some of these factors and will present a few general guidelines to help in the selection process.

FACTORS TO CONSIDER

Planning for a statistical analysis of a study's data begins very early in the research process, at about the same time that decisions are made about what measuring instrument(s) is going to be used to measure certain variables of interest. Questions relating to how variables should be measured and how they will be best statistically analyzed are closely related. The way variables are measured helps to determine which statistical tests can and cannot be used to test a research hypothesis.

It is customary to select and specify the statistical test(s) that is going to be used *prior* to data collection. However, it is not unusual to encounter problems in a research study that may change either the way data are collected and/or the way in which variables can be measured. When this occurs, it is considered ethical and, in some cases, absolutely essential to select statistical tests different from those stated in the original research design.

Whenever the final choice of one or more statistical tests is made, three considerations most directly influence the choice: (a) the sampling method(s) used, (b) the nature of the distribution of the dependent variable (and sometimes the independent variable as well) within the research population, and (c) the level of measurement of the independent and dependent variables.

Sampling Methods Used

The characteristics of a sample and the way that it is drawn from a population influence the choice of a statistical test. Certain tests require certain types of sampling methods.

The topic of sampling is usually covered in detail in most social work research methods texts. For our purposes, we will just emphasize that researchers must be fully aware of what sampling methods they used if they want to select the most appropriate tests to test their research hypotheses. Below are some of the questions they must address:

1. Did the sampling method use a single sample or more than one sample? How many?
2. If more than one sample was used, were the samples independent of each other or were they related in any way? For example, were they "matched" in regard to certain variables to assure that they were comparable?
3. Were the cases within the samples drawn "independently"? For example, did the selection of one case in any way increase or decrease the likelihood that any other case within the population would also be selected?
4. Was a random or "probability" sampling method used? Did we know the probability of every case being drawn? Was that probability the same for every other case?

Answering these questions allows the researcher to narrow the search for the appropriate statistical test. It will eliminate a majority of the existing tests because of their inappropriateness for the particular sampling method that was employed, but it is only one step in the "process of elimination."

Nature of the Research Population

A second major consideration in selecting a statistical test is the way in which the dependent variable is distributed within the population from which the research sample was drawn. Some of the most useful tests require that this assumption is met. As we have seen in Chapter 3, a positively or negatively skewed distribution of the dependent variable usually precludes the use of the mean as a measure of central tendency or the standard deviation as a measure of variability. Without a normal distribution of a dependent variable, many potentially useful tests that contain the mean and/or standard deviation in their formulas must be eliminated from consideration.

As we noted in Chapter 4, normal distributions are rarely perfect in their symmetry. Many times, full descriptive data on a given dependent variable within a population do not exist; if they do, the data may only approximate a bell-shaped curve. The decision about whether the distribution of a dependent variable within a population is sufficiently "normal" to be regarded as normal is often a "judgment call."

Generally, a dependent variable whose values at least approximate a bell-shaped curve within the population would justify the use of certain statistical tests that require a normal distribution of the variable. Judgments of this type are common and necessary in statistical analyses. For example, we make a similar judgment when we decide (a) when enough literature has been reviewed, (b) that our sample is sufficiently large and representative of the population from which it was drawn, (c) that we can or cannot justify the use of a one-tailed hypothesis, or (d) what we believe to be an appropriate rejection level for the null hypothesis.

Level of Measurement

A third major factor to consider in the selection of a statistical test is the level of measurement of the independent and dependent variables. Are the variables regarded as nominal, ordinal, interval, or ratio? Well-planned construction and use of measuring instruments help us to generate the highest possible level of measurement for any given variable. However, through carelessness, we can throw away data precision. For instance, we may permit a variable that could have been measured at the interval or ratio level to be measured at only the nominal or ordinal level. Once measurements have been made, it is impossible to achieve the higher level of measurement precision than was initially possible.

Why is loss of measurement precision important? After all, are there not different statistical tests designed for use with different levels of measurement? So what difference does it make what level of measurement is used? In fact, the distinction between interval and ratio level measurement is relatively unimportant in the selection of statistical tests. Many statistics books simply use the term "interval/ratio" to note the presence of either level of measurement, but the presence of only nominal or ordinal measurement when interval or ratio is possible can result in a real loss to the researcher.

The use of a measurement that yields a lower level of measurement (i.e., nominal or ordinal) automatically precludes the use of all statistical tests that require interval or ratio level variables. This would not be problematic except, as we shall see, those tests that require interval or ratio measurement of one or more variables are some of the best tests developed for identifying true relationships between and among variables. They are also some of the most powerful tests.

THE CONCEPT OF STATISTICAL POWER

In selecting a statistical test, the concept of statistical power must be considered. *Statistical power* is the ability of a test to correctly reject the null hypothesis; that is, its ability to correctly detect a true relationship between or among variables.

Based on their mathematical computations, some statistical tests are inherently more powerful than others; that is, some are better able to detect a true relationship between or among variables. Some may allow a researcher to justify rejection of the null hypothesis, while other, less powerful tests would not. The more powerful tests usually have demanding conditions for their use. These conditions must be met in order for the tests to be used correctly.

As presented in Chapter 3, the standard deviation, where appropriate, is preferable to the range as an indicator of dispersion and the mean, where appropriate, is a more precise indicator of central tendency than is the median or mode. Why? Both the mean and the standard deviation require computations using every value for the variable's distribution, whereas the less precise descriptive statistics do not.

The same principle applies in understanding the power of statistical tests that are used in hypothesis testing. More powerful tests use the actual values for all cases in their computations (directly or indirectly) rather than, for example, the ranks assigned to cases or the frequencies for different nominal level categories of variables. Thus, they take full advantage of the greater precision in measurement that is available. Not surprisingly, the formulas for more powerful statistical tests also tend to be more complex than the formulas for less powerful ones.

Before selecting a statistical test, we need to determine how much power is desirable (and appropriate) for examining the predicted relationship between two variables. We need to know what would constitute "not powerful enough" or "too powerful."

Having decided what we need in a test, we can explore available alternatives. We may find one that appears to be about what we need, but it may not seem quite powerful enough or a little too powerful, which may not be a problem. All tests possess a certain inherent power, based upon their mathematical formulas, but any test can be made more or less powerful by conditions related to its use.

Factors That Affect Statistical Power

What makes a statistical test more or less able to detect a true relationship between two variables in a given data set? Three conditions affect the statistical power of a given test: (a) the strength of the actual relationship between two variables that exists within the population, (b) the predetermined statistical rejection level (e.g., .05, .01, .001) that is used with the test, and (c) the size of the research sample used.

We cannot influence the strength of the actual relationship between two variables; it generally is regarded as a "given" that we would like to know more about. Very strong relationships are likely to be detected by almost any test that is used, but relatively weak ones may be detected only by more powerful tests. However, we *can* affect the statistical power of a test by influencing either one or both of the other two conditions. At the beginning of a study, we can select either a higher or lower rejection level than the conventional .05 and/or increase or decrease the study's sample size.

If, for example, we decide to use the rejection level of .10 as the cutoff point at which we will reject the null hypothesis, we can increase the likelihood of not "missing" a true relationship between two variables that exists in the population and, thus, of not committing a Type II error. However, as noted in Chapter 5, this will increase the likelihood of committing a Type I error.

On the other hand, if we were to go in the opposite direction and use the .01 level, we would make any given statistical test less powerful, but we would decrease the likelihood of concluding that two variables are related when they really are not (Type I error). In which direction, if either, should we go? It depends on the consequences of making either error and which would be the more tolerable from an ethical and practice perspective. We would want to make a test either more powerful or less powerful only after a careful assessment of the practice implications of such a decision.

Consideration of sample size, along with the ethical and practice considerations, allows the researcher to select a test that will generate the optimal amount of statistical power. There usually is a sample size that is considered ideal for use with a given statistical test. It is specified in advanced statistics books that discuss various tests and their appropriate usage. However, selecting a larger sample than recommended will make any statistical test more powerful, and selecting a smaller sample will make any test less powerful.

The fact that the inherent statistical power of a test can be adjusted through sample size is an important point to remember. Sample selection, if not carefully planned, may result in a statistical analysis that is not powerful enough to identify a true relationship between two variables. It can also produce an analysis that is too powerful, however, resulting in conclusions about relationships that can become dangerously misleading to practitioners and other researchers. An overly powerful statistical analysis can detect relationships between two variables that may be real but are so weak in the absolute sense that they are best left undiscovered and unreported. We will return to this point in later chapters.

PARAMETRIC AND NONPARAMETRIC TESTS

There are two general types of statistical tests—parametric and nonparametric. *Parametric* tests are more powerful than nonparametric tests and generally more desirable, if conditions for their use can be met. Parametric tests generally require:

(a) a normal distribution of scores within the population being studied, (b) the drawing of independent samples, and (c) at least one variable that is either at the interval or ratio level of measurement. Additional requirements apply to certain parametric tests. Generally, it is helpful to remember that if the mean and the standard deviation are appropriate descriptive statistics for summarizing a study's findings, parametric statistics *may* be appropriate for examining the relationships between or among variables within the study.

Nonparametric tests are designed for research situations in which the conditions for the use of parametric tests do not exist. They are less powerful as a group than parametric tests, and they do not require a normal distribution of the dependent variable within the population from which the research sample was drawn. Some are intended for independent samples; others are not. The number of samples and the number of cases within each sample also are important factors in selecting one specific test from the many that exist. Most nonparametric tests require only nominal or ordinal level data, but some are more demanding, requiring greater measurement precision.

Because nonparametric tests generally are designed for the analysis of nominal or ordinal level data that need not be normally distributed, they are often ideally suited for social work research. As noted earlier, many of the dependent variables of common interest to social workers are not at the interval or ratio levels (e.g., success or failure in treatment, rehospitalization or non-rehospitalization, passage or nonpassage of legislation).

Nonparametric tests are more than just a second best choice designed for situations in which criteria for parametric statistics cannot be met. They have some distinct advantages over parametric tests and often are the best test for addressing some of our statistical needs. For example, a nonparametric statistic is especially useful when:

1. Samples have been compiled from different populations.
2. Data measurement consists primarily of rank ordering of several responses to a question.
3. Very small samples (as few as six or seven) are all that are available for study.

Fortunately, the relative lack of power of nonparametric tests can be compensated for, at least in part. As discussed above, increasing sample size can increase the power of any statistical analysis. In many situations where nonparametric tests must be employed, two or more tests could potentially be used, but they are likely to have different sample size requirements. As a general rule, the test requiring the largest sample size is likely to be the most powerful one. If we anticipate the need for more power in testing (such as when we anticipate that a relationship between variables may be weak, but we wish to document its existence), we always can consider the possibility of using a larger sample(s). This would allow us to justify the use of the more powerful statistical test. Many nonparametric tests can be made just about as powerful as their parametric counterparts with the use of sufficiently large samples.

GENERAL GUIDELINES FOR TEST SELECTION

Generally, the most powerful test that can be justified for any given research situation should be used. Data are wasted if a less powerful test is used when the criteria for a more powerful test can be met. However, a test should only be used under the conditions for which it was intended. We can avoid the use of either a test that is not powerful enough or one that is too powerful by adequately planning for statistical analysis when the operationalization of variables is being considered and when sampling decisions are made, and by becoming knowledgeable about various tests and their appropriate usage.

Figure 6.1 presents a guide for selecting the most appropriate statistical test when the relationship between one independent variable and one dependent variable is examined. The level of measurement for the dependent variable guides us in choosing a parametric or a nonparametric test and in deciding what particular test can be used. Figure 6.1 displays the names of several parametric tests (lefthand side) discussed in this book and their corresponding nonparametric alternatives (righthand side).

Although parametric tests traditionally are believed to require interval or ratio level data for the dependent variable, some statisticians argue that ordinal level data can be used when the intervals on the ordinal scale are approximately equal. The decision to use ordinal level data with parametric statistics involves a close review of the data.

What is certain is that a nominal level dependent variable can only be assessed using nonparametric statistical tests. Consequently, there are no parametric alternatives for Chi-square, Fisher's Exact test, and McNemar's test in Figure 6.1. Nonparametric tests (to be discussed in Chapter 11) are most frequently used when the sample size is small, the sample distribution for the dependent variable is badly skewed, or the data do not meet other assumptions of parametric tests.

The level of measurement for the independent variable assists us in choosing a specific statistical test (see Figure 6.1). With different levels or types of data for the independent variable, particular parametric and nonparametric tests are appropriate. There are a number of comprehensive source books on statistical test selection, many of which are available in most university libraries. Some present highly detailed flow charts for use in selecting statistical tests given various assumptions about a study's research design and the level of measurement of variables. To use such reference books, it is essential to know: (a) the level of measurement of the dependent variable, (b) the level of measurement of the independent variable, (c) exactly how the research sample was obtained and its size, and (d) how the dependent variable (and independent variable, if appropriate) is distributed within the population from which it was drawn.

FIGURE 6.1 Using level of measurement to determine the appropriate statistical test

Parametric Statistical Tests

Test	Dependent (Criterion) Variable	Independent (Predictor) Variable
—	—	—
—	—	—
—	—	—
Group t-test	Interval/ratio	Nominal (2 categories)
Correlated t-test	Interval/ratio	(2 repeated measures)
One-way ANOVA	Interval/ratio	Nominal (3 or more categories)
Within subject one-way ANOVA	Interval/ratio	(2 or more repeated measures)
Pearson's r	Interval/ratio	Interval/ratio

Nonparametric Statistical Tests

Test	Dependent (Criterion) Variable	Independent (Predictor) Variable
Chi-square test	Nominal (2 or more categories)	Nominal (2 or more categories)
Fisher's exact test	Nominal (2 categories)	Nominal (2 categories)
McNemar's test	Nominal (2 categories)	(2 repeated measures)
Mann-Whitney U test	Ordinal	Nominal (2 categories)
Wilcoxon Sign test	Ordinal	(2 repeated measures)
Kruskal-Wallis test	Ordinal	Nominal (3 or more categories)
Friedman's test	Ordinal	(2 or more repeated measures)
Spearman rho test Kendall's tau test	Ordinal Ordinal	Ordinal Ordinal

OTHER SOURCES OF ASSISTANCE

This chapter has focused on the process involved in selecting statistical tests to analyze the relationship between or among variables. It has introduced the reader to the complicated task of selecting the most appropriate statistical test for hypothesis testing and noted some of the many different factors that need to be considered. We have definitely not provided the answer to the question, "What statistical test should *I* use given *my* specific research situation?" While Figure 6.1 or the books listed in the References and Further Readings section located at the end of this book can be helpful in test selection, computer technology has now made it even easier for us to select a test.

Computer statistical software packages are now available that use the information provided—the level of measurement of the independent and dependent variables and the shape of their distribution and the sampling method used—to suggest which appropriate statistical test(s) to use. As these programs become more widely available, decision-making charts such as Figure 6.1 will no longer be necessary. (In fact, the authors of one of the most comprehensive test selection reference books now have a computer version of their book.)

Computers can tell us only what we have "statistically found" in our data. Even with computer-assisted statistical test selection and high-powered number crunching, we still need to know how to interpret the findings from a computer-generated statistical analysis and to be able to relate them to social work practice. Computers also do not tell us (not yet, anyway) what meaningful research problems to pursue or which hypotheses to test. They cannot tell us how to conceptualize, operationalize, design, implement, and gather data to test our research hypothesis. For now, anyway, these tasks are left up to social work researchers and practitioners.

For social workers who lack access to either software packages for computer-assisted statistical test selection and/or to the most up-to-date reference books, consultation with experts is available to assist in the task of research design and statistical analyses. In schools and departments of social work and many human service organizations, experts can offer good advice. They can be especially helpful if we know what questions to ask and how to ask them, and if we can comprehend the questions they ask of us.

Because so many statistical tests are available, even the experts cannot be knowledgeable about all of them. In fact, probably fewer than a dozen tests are commonly used in social work research, another group is seen occasionally in the literature, and a third group consists of a large number of obscure tests. New tests are being created constantly with the assistance of computers, while others that were popular for a time have dropped out of favor.

The chapters that follow present a description of a relatively few tests that have been widely used in social work research for many years and promise to remain popular. They are both versatile and suitable for many types of data analysis situations found in the social work profession. The emphasis will be on understanding what they do and what their results mean rather than on the mathematics of their computations. In today's professional environment, we usually can leave that job to the computers.

CONCLUDING THOUGHTS

This chapter and Chapter 5 have focused on one of the most critical components of the research process—hypothesis testing. Understanding its theoretical underpinnings and knowing just how it works make it possible for us to comprehend why statistical analysis is far more than just number crunching. We can now summarize the process of hypothesis testing as a series of steps to be completed:

1. *State the research hypothesis or hypotheses.* Is each considered to be one tailed, two tailed, or in the null form? What are the independent (predictor) and dependent (criterion) variables? How is each conceptualized and operationalized?

   Remember: The research hypothesis guides the entire conceptualization and operationalization process. In addition, it helps to determine the best research design to use in an effort to rule out alternative explanations that could explain an apparent relationship between or among variables. Research hypotheses should be supported by theoretical assumptions derived from a literature review.

   A research hypothesis can be one or two tailed (and sometimes it can be in the null form). The research hypothesis is not statistically tested; it is either indirectly supported or not supported by directly testing the null hypothesis.

2. *For any one-tailed or two-tailed hypothesis, state the corresponding null hypothesis.* How would the data look if we were unable to reject the null hypothesis?

   Remember: The null hypothesis states that there is no relationship between or among the variables in the research hypothesis. The null hypothesis covers all outcome possibilities that are not explicit in the research hypothesis. For example, if a one-tailed research hypothesis states that the level of job satisfaction for medical social workers has increased significantly over the last four years, the null hypothesis must include the possibilities that it has either decreased or that it has not changed.

   If the null hypothesis can be rejected, support is present for the research hypothesis. Hypothesis testing is a process of indirect proof. We never directly prove that the research hypothesis is correct; rather, if the null hypothesis can be rejected, the research hypothesis (the alternative hypothesis) is supported indirectly.

3. *Specify the statistical rejection level to be used.* If any level other than .05 is to be used, what is the justification for its use? Which type of error, Type I or Type II, are we seeking to avoid and why?

Remember: This step entails determining the consequences of erroneously rejecting the null hypothesis within our research situation. Most social work researchers prefer to take a conservative position by setting a low rejection level (e.g., .01, .025, .05) rather than a higher one (e.g., .10, .15). They would rather chance committing a Type II error than committing a Type I error.

4. *State any assumptions about the data and how they were collected.* What levels of measurement are assumed to exist? Which variables are assumed to be normally distributed? What specific sampling methods were employed? How large were the samples?

   Remember: We must know the level of measurement for each of the variables that we want to analyze. The method by which the sample was drawn, sample size, and the way our variables are distributed in the population also affect the choice of which statistical test to use.

5. *Select and compute one or more test statistics.* Is each test used appropriate for the conditions described in step 4? Is a computer package or consultant to be used in the selection of the tests? Is computation to be computer assisted? If so, what statistical software package will be used? Conceptually, how will each test generate a *p* value?

   Remember: The statistical test used does not determine the degree to which rival hypotheses, design bias, and sampling bias helped to create an apparent relationship between or among variables. It only determines the likelihood that chance played a role in the study's findings.

6. *Determine whether the relationship between the variables is statistically significant.* Is the probability (*p* value) smaller than the predetermined rejection level? If so, are we reasonably certain that some other factor besides chance (rival hypotheses, design bias, sampling bias) did not produce the relationship between the two variables? If so, it probably is safe to reject the null hypotheses.

   Remember: We can never be totally certain that chance did not produce an apparent relationship between or among variables. When we reject the null hypothesis, we are saying that we are reasonably certain that the apparent relationship is a real one—but we might still be wrong!

7. *Determine if each statistically significant relationship is meaningful.* To what degree did sample size contribute to statistical significance? How strong is the absolute relationship between the two variables? How useful are the study's findings to the social work practitioner, educator, or researcher? To what extent would we feel safe in generalizing the findings beyond our sample (external validity)?

   Remember: These decisions require a thoughtful combination of ethics, common sense, convention, and practice expertise. They must be addressed before a study's findings are implemented in the social work practice milieu.

STUDY QUESTIONS

1. How can the use of an inappropriate statistical test harm the credibility of the research study?

2. How might a researcher's use of an inappropriate test ultimately have a negative effect on services to clients? Discuss. Provide a social work example in your discussion.

3. How does the "rule of the instrument" sometimes lead to the selection of an inappropriate statistical test?

4. What factors related to sampling methods used help to determine which statistical test is appropriate?

5. What two characteristics of the data also affect which statistical test should be used?

6. How does the operationalization of a variable performed before data are even collected serve to limit (or expand) our option of a statistical test that can be used? Provide original examples in your discussion.

7. What do we mean when we say that one statistical test is more "powerful" than another?

8. How is it possible for a test to be "too powerful"? What problems can a too powerful test create?

9. How is it possible for a test to be "not powerful enough"? What problems can a not powerful enough test create?

10. What are two ways to adjust the power of a test?

11. What three criteria must be met for a parametric test to be used?

12. Why are nonparametric tests particularly useful in many social work research projects? How can we make them more powerful?

chapter 7

Cross-Tabulation

There is no better choice than to begin the study of inferential statistics with cross-tabulation and a related statistical test known as chi-square. Cross-tabulation and chi-square are both easily understood. *Cross-tabulation,* as we are using the term here, refers to the process of putting the values of two nominal level variables into a simple table. The table consists of the number of times that each possible combination of the values for the two variables occurred within the research sample. If cross-tabulation can be performed, it is possible to test for statistical significance of the relationship between the variables using chi-square.

THE POPULARITY OF CHI-SQUARE

No nonparametric statistical test is better known than chi-square. Most social workers who have been exposed to statistical analysis as part of their professional social work education have spent some time studying it. Research articles using chi-square analyses appear frequently in the professional social work literature. Social workers feel comfortable with the statistic, and are more likely to read reports of research studies that used it than research studies that employed less widely known and understood statistical tests.

A major reason for the popularity of the chi-square test is that it requires only nominal level data, which means the values for each variable need only represent distinct categories and a difference of kind. This fact alone makes chi-square especially well suited for a large number of social work research situations. For example, social workers often want to know if there is a relationship between their efforts (the independent variable) and their clients' outcomes (the dependent variable).

When measuring treatment effectiveness, we frequently can do little more than assign cases to categories, such as *found employment/did not find employment, abused again/did not abuse again, rehospitalized/not rehospitalized—* nominal level measurement. Other variables that may be related to treatment effectiveness also are likely to be at the nominal level of measurement. These variables could be, for example, *the type of treatment provided* (e.g., individual or group treatment), *the professional background of the social worker* (e.g., B.S.W. or M.S.W.), or *the demographic variables of the client* (e.g., marital status, gender, religious affiliation, race).

Similarly, many other variables in a social work research study have *yes-no* or other dichotomous value categories that can be considered to be only at the nominal level of measurement. For example, a small research study might measure whether a piece of social legislation *passed or did not pass* in different states (dependent variable) and *whether or not* the local chapters of the National Association of Social Workers supported the legislation (independent variable). On the other hand, a study could examine whether there is a relationship between whether candidates for local offices *were elected or not* (dependent variable) and *whether or not* they had taken a pro-choice position on the issue of abortion. In all these situations, the variables are at the nominal level. Because of this fact, chi-square can be the ideal statistical test to use. However, as we shall see, chi-square is not appropriate for all data analyses situations involving two nominal level variables.

UNDERSTANDING CHI-SQUARE

Chi-square is a statistical test of association between variables. Even if we can demonstrate support for a relationship between the variables by using it, it would not be appropriate to conclude that the different values of either variable *caused* the different values of the other. We can only state, at best, that a pattern or clustering of values may exist—certain values of one variable tend to be found where certain values of the other value are also present. Like all statistical tests, chi-square can determine whether any pattern within the data for a research sample is so strong and consistent that chance is an inadequate explanation of it.

As noted in Chapter 5, all statistical tests attempt to refute chance as one explanation for an apparent relationship between two variables, resulting in their ability to reject the null hypothesis. With chi-square, the specter of chance plays the skeptic in a unique way; its role is central to understanding this very useful test.

Let us use a simple hypothetical example to illustrate how chi-square works. Suppose that we want to find out if there is a relationship between a nominal level dependent variable, such as *client outcome* (success/failure), and a nominal level independent variable, such as *type of treatment* (group/individual), within a large alcoholism counseling program. Based on a literature review and previous experience and hunches, we feel justified in formulating a one-tailed

TABLE 7.1 Dummy table: Type of treatment by client outcome

Type of Treatment	Results		Totals
	Success	Failure	
Group	*		
Individual		*	

research hypothesis: "Clients in group treatment are more likely to abstain from alcohol for two months than clients in individual treatment." When a one-tailed hypothesis such as this one is stated, and we are planning to use the chi-square statistic to test the hypothesis, it is extremely helpful to depict the hypothesis in the form of a simple table, referred to as a *dummy table*.

A dummy table shows where we would expect to find a disproportionately large number of cases if our one-tailed research hypothesis is supported through statistical testing. In our example, if the research hypothesis is supported, we would expect to find a relatively large number of cases who were in group treatment who abstained from alcohol (success) and a relatively large number of cases who were in individual treatment who did not abstain (failure). This is seen by the placement of the asterisks in the body of Table 7.1. As we shall soon see, the creation of a dummy table at the time that a one-tailed hypothesis is formulated can assist greatly in the interpretation of a study's findings.

As emphasized in Chapter 5, the concept of the null hypothesis always guides our decision-making process when we are testing our research hypothesis. The corresponding null hypothesis for the one-tailed research hypothesis would state, "There is no relationship between whether or not clients abstained from alcohol for two months and the type of treatment they received." The null hypothesis would assert that, although it may occasionally seem (within a research sample) that successful treatment is more likely to occur among clients who received group treatment than among those who received individual treatment, there is not a real relationship between the two variables within the population from which the sample was drawn.

To test our research hypothesis, 100 current cases could be randomly selected from the program's 300 alcoholic clients. These 100 clients would form the research sample. We could ask the social workers to note for each client they saw (a) whether the client was in individual or group treatment and (b) whether or not the client abstained from alcohol for the previous two-month period.

Type of treatment would have two possible values—group or individual. Client outcome also would have two possible values—success or failure. Success would be operationalized as two months of alcohol abstention. Thus, there are four possible "combinations" of values that could be recorded for each client, with every client falling into one and only one of the four categories:

- group treatment *and* client success
- group treatment *and* client failure
- individual treatment *and* client success
- individual treatment *and* client failure

Suppose that the 100 clients were distributed among each of the four categories as follows:

- group treatment *and* client success (40 clients)
- group treatment *and* client failure (20 clients)
- individual treatment *and* client success (15 clients)
- individual treatment *and* client failure (25 clients)

Observed Frequencies

These data can now be placed into a *cross-tabulation table,* which will contain the actual data from the research study. The data in the four cells of Table 7.2 are the actual number of clients who possessed each combination of values for the two variables. Such a table (also referred to as a chi-square table, cross-tab, cross-break, or contingency table) is constructed using the variable labels, value labels, and frequencies drawn from the data. A cross-tabulation table for our hypothetical data is presented as Table 7.2.

In the far righthand column of Table 7.2, the totals for each row are entered. Likewise, the totals for each column (adding down) are entered on the bottom line. These row and column totals are called *marginal totals,* or marginals. They indicate the total number of cases that were observed to possess a given value for each variable: group treatment (60), individual treatment (40), success (55), and failure (45). The grand total of cases (*N*) in the table is entered in the bottom righthand corner (100). Note that the sum of the row totals equals the grand total, as does the sum of the column totals.

The areas where the frequencies (40, 20, 15, and 25) are placed are referred to as "cells." Table 7.2 contains two dichotomous (two-category) variables, *type of treatment* and *client outcome;* thus, it has four cells. It is, of course, possible

TABLE 7.2   Observed frequencies: Type of treatment by client outcome

Type of Treatment	Results		Totals
	Success	Failure	
Group	40 (*a*)	20 (*b*)	60
Individual	15 (*c*)	25 (*d*)	40
Totals	55	45	100

to have variables with more than two value categories. The cross-tabulation table would then have more rows and columns and thus more cells.

By convention, small italicized letters (*a, b, c, d,* etc.) are used to refer to the respective cells. They are lettered left to right, starting with the top row, then going to the next row, and so on.

The variables are nominal; no rank ordering or quantitative difference is implied by their value labels. Consequently, there is no logical sequence in which they should appear in a table. In our example, it would have been equally correct to place *individual treatment* above *group treatment* or to place *failure* to the left of *success*. However, by convention, we usually display the dependent variable in the columns and the independent variable in the rows, as seen in Table 7.2. If neither variable is clearly independent or dependent, it makes no difference which variable's value categories run down the side of the table and which run across its top.

Table 7.2 displays what is referred to as the actual or *observed frequencies* derived from the data collected in our hypothetical study comparing the two treatment methods for alcoholism. Now that our data have been placed in the cross-tabulation table, we can compare the outcomes of clients who received group treatment with outcomes of those who received individual treatment. Such a comparison cannot be easily accomplished using only Table 7.2.

We cannot simply compare the number of clients who had successful outcomes with group treatment (40) directly with the number of clients who had successful outcomes with individual treatment (15) and conclude that group treatment is better because 40 is larger than 15. We would naturally expect to have more successes among those clients in group treatment than among those in individual treatment because more clients (60) were in group treatment and fewer (40) were in individual treatment.

It is possible to compensate for the difference in the number of cases in the two subsamples (individual versus group) by using percentages. They "equalize" the sizes of the two groups. We can determine what percentage 40 clients is of 60 clients and what percentage 15 clients is of 40 clients. Table 7.3 is a percentage table for the observed data contained in Table 7.2. It indicates that 66.7 percent of the 60 clients who received group treatment had successful outcomes, compared with 37.5 percent of the 40 clients who received individual treatment and had successful outcomes.

TABLE 7.3 Percentages of clients: Type of treatment by client outcome

Type of Treatment	Results		Totals
	Success	Failure	
Group	66.7	33.3	100.0
Individual	37.5	62.5	100.0

If the percentages in cells *a* and *c* and in cells *b* and *d* had been exactly the same, we would have no reason to believe that the two variables are related. However, the percentages in the cells in Table 7.3 are dramatically different. They seem to suggest that, among alcoholic clients, type of treatment and treatment success *may* be related. Yet, at this point, it is difficult to know whether the apparent relationship is anything more than the work of chance. The null hypothesis would argue that a 29.2 percentage point difference (66.7 - 37.5 = 29.2 points) is not really very large. But is it? How much of a percentage difference is needed to rule out chance as an apparent relationship between the two variables? Percentages alone cannot answer this question.

Expected Frequencies

In order to answer the previous question, we need to introduce the concept of expected frequencies. *Expected frequencies* are those frequencies that we would expect to occur most frequently if the null hypothesis were correct—that is, if there was not a real relationship between the two variables. Unlike observed frequencies, which reflect actual data that we collected (as in Table 7.2), expected frequencies are hypothetical numbers; they are derived using simple math from the marginal totals in a cross-tabulation table.

To understand expected frequencies, return to Table 7.2. Let us see how the expected frequencies for a given cell are obtained. Look first at cell *a*, which contains the 40 clients who were in group treatment *and* who had successful client outcomes. They represent two thirds, or 66.7 percent, of the 60 clients who were in group treatment (Table 7.3).

The expected frequency for cell *a* (or for any cell) is based upon the marginal totals that correspond to that cell; that is, the total for the row in which that cell appears and the total for the column in which that cell appears. For cell *a* in Table 7.2, these two marginal totals are 60 and 55, respectively. For cell *b*, they are 60 and 45, respectively. For cell *c*, they are 40 for the row total and 55 for the column total, and for cell *d*, these are 40 and 45.

The expected frequency for any given cell is computed by multiplying that cell's row total by its column total and then dividing by the grand total for the table (*N*). We can express this as a simple formula:

$$E = \frac{(R)\ (C)}{(N)}$$

Where:

E = Expected frequency in a particular cell
R = Marginal total for the row in which the cell appears
C = Marginal total for the column in which the cell appears
N = Total number of cases

TABLE 7.4 Observed and (expected) frequencies:
Type of treatment by client outcome

| Type of Treatment | Results | | Totals |
	Success	Failure	
Group	40 (33)	20 (27)	60
Individual	15 (22)	25 (18)	40
Totals	55	45	100

Using the formula, we note that the expected frequency for cell a would be $60 \times 55 = 3300/100 = 33$; for cell b it would be $60 \times 45 = 2700/100 = 27$; for cell c it would be $55 \times 40 = 2200/100 = 22$; and for cell d it would be $45 \times 40 = 1800/100 = 18$. Table 7.4 displays the expected frequency for each of the four cells in parentheses alongside the corresponding observed frequency. Note that there is only one set of marginal totals, because they are constant for both the observed frequencies and the expected frequencies.

Chi-square compares the observed frequency for each cell with its respective expected frequency. It uses a formula to produce a chi-square value which is directly related to the size of these differences. Larger differences between the observed and expected frequencies produce larger chi-square values. Let us now turn to how a specific chi-square value is calculated using Table 7.4.

COMPUTATION OF CHI-SQUARE

In the not too distant past, calculating a chi-square statistic (or any statistical test for that matter) was a time-consuming task. It is now a rare occasion when a chi-square analysis is undertaken with pencil and paper. When it is, there are seven steps involved:

1. Place the observed frequencies into a cross-tabulation table.
2. Compute the expected frequencies and add them to the table.
3. Compare the observed frequencies (Step 1) with the expected frequencies (Step 2) using the chi-square formula.
4. Compute degrees of freedom (to be explained later) for the cross-tabulation table.
5. Go to the appropriate line for the degrees of freedom in a critical values of chi-square table (see Appendix B) and see where the chi-square value falls.
6. Go to the number to the left of where the chi-square value falls.
7. Go to the top of the column in which that number appears and obtain the approximate p value using the line corresponding to the type of research hypothesis used (one tailed or two tailed).

The general formula that is needed to compute a chi-square statistic is contained in Box 7.1. It is used for cross-tabulation tables containing more than four cells. If there are only four cells, as is the case in our example, we would normally adjust the formula slightly by reducing the absolute difference between the observed and expected frequencies by .5 for each cell before squaring it (referred to as the *Yates Correction Factor* or the correction for continuity). This is a mathematical correction designed to avoid the slightly inflated chi-square value that the uncorrected general formula will produce in tables that have only four cells (referred to as 2 × 2 tables). For the sake of simplicity (and because most cross-tabulation tables contain more than four cells), the computation illustrated in Box 7.1 does not include the use of the Yates Correction Factor. However, other chi-square values displayed in the tables in this chapter were derived from the corrected formula where appropriate.

Today any statistical software program that has the capacity to perform a chi-square analysis will save us a lot of time. The formula for chi-square along with the formulas for expected frequencies (previously discussed) and degrees of freedom (to be discussed) are all preprogrammed. Using an inexpensive statistical software package, we could enter the data for our 100 cases in the example in this chapter in just a few minutes. We could then instruct the computer to run a chi-square test to examine the association between type of treatment and client outcome. In less than two seconds, all formulas will have been computed and the output of the analysis will appear on the screen.

In our example, we would find that the chi-square value for the data contained in Table 7.2 is 8.24 without using the Yates Correction Factor, and that it is reduced to 7.11 when the Yates formula is used. We would also find that there was one degree of freedom, and the p value for the data, using either formula, was "less than .005." Almost instantly, we would be ready to interpret the results of the analysis.

Relating Chi-Square to Chance

If we were to study the chi-square formula in Box 7.1 closely, we could see why larger differences between observed and expected frequencies generate larger chi-square values. Generally, the larger the chi-square value, the less likelihood that chance caused the differences between the observed and expected frequencies. This is seen in tables of critical values of chi-square, such as the one in Appendix B. We can easily compute a chi-square value for our data in Table 7.2. However, before we could use Appendix B to determine whether a statistically significant association exists between the two variables, we would first need to understand and calculate degrees of freedom.

Degrees of Freedom. If we were to study the chi-square formula a little more, we could see why larger differences between observed and expected frequencies are not the only thing that will generate larger chi-square values. The likelihood

BOX 7.1
The Chi-Square Statistic
(without Yates Correction Factor)

Hypothesis: Clients who received group treatment are more likely to abstain from alcohol than those clients who received individual treatment.

Null Hypothesis: There is no difference between the type of treatment clients received and whether they abstained from alcohol.

Predictor Variable: Type of treatment—nominal level (group versus individual).

Criterion Variable: Client outcome—nominal level (success versus failure).

Observed Frequencies: See Table 7.2.

Expected Frequencies Formula: $E = \dfrac{(R)\,(C)}{(N)}$

Where:

E = Expected frequency in a particular cell C = Total number in that cell's column
R = Total number in that cell's row N = Total number of cases

Substituting values for letters:

Cell a: $E = \dfrac{(60)\,(55)}{100} = 33$ Cell c: $E = \dfrac{(40)\,(55)}{100} = 22$

Cell b: $E = \dfrac{(60)\,(45)}{100} = 27$ Cell d: $E = \dfrac{(40)\,(45)}{100} = 18$

Chi-Square Formula: $x^2 = \Sigma\,\dfrac{(O-E)^2}{E}$

Where: x^2 = Chi-square value E = Expected frequency
         O = Observed frequency Σ = Sum of (for all cells)

Substituting values for letters:

$$
\begin{array}{ccccccc}
\text{Cell } a & + & \text{Cell } b & + & \text{Cell } c & + & \text{Cell } d \\
x^2 = (40{-}33)^2 & + & (20{-}27)^2 & + & (15{-}22)^2 & + & (25{-}18)^2 \\
= 8.24 \text{ (Chi-square value)}
\end{array}
$$

Degrees of Freedom Formula: $df = (r-1)\,(c-1)$

Where: df = Degrees of freedom
        r = Number of rows
        c = Number of columns

Substituting values for letters:

df = $(2-1)\,(2-1)$
    = 1 (Degree of freedom)

Presentation of Results: $x^2 = 8.24$, $df = 1$, $p < .005$

Conclusions: The null hypothesis is rejected and the one-tailed research hypothesis is supported. In short, clients who received group treatment had statistically significant better outcomes than clients who received individual treatment.

of obtaining a large chi-square value is also affected by the size of the cross-tabulation table on which it is computed. As used here, size refers to the number of rows and columns and, more specifically, to the number of cells in the table.

The larger the table, the more likely it is to have a large chi-square value. This should be evident by glancing at Box 7.1, as the chi-square value is the sum of figures derived from comparison of observed and expected frequencies within each of the cells. Thus, the more cells in a table, the more cells available to contribute to the chi-square value and the higher that value is likely to be. This will occur whether the actual differences between observed and expected frequencies are large or relatively small.

A chi-square value must be evaluated in relation to the size of the table from which it was computed. A large chi-square value in itself may not mean a statistically significant relationship between two variables. It could have been generated by a large table with many cells, each contributing a small amount (because of small differences) to that value. Conversely, a smaller chi-square value may be statistically significant if the table from which it was produced had relatively few cells to contribute to its value. The critical values of chi-square contained in Appendix B reflect this.

The number of cells in a cross-tabulation table is expressed in terms of degrees of freedom. The degrees of freedom for any chi-square analysis are equal to the number of rows minus one times the number of columns minus one (see Box 7.1). Using this formula tells us that Table 7.2 has one degree of freedom (as will all 2-by-2 tables). A 3-by-3 table would have four degrees of freedom $(3 - 1) \times (3 - 1) = 4$, a 2-by-3 table would have two degrees of freedom $(2 - 1) \times (2 - 1) = 2$, and so on.

Restrictions on Use of Chi-Square

It is helpful to understand what an expected frequency is and where it comes from. How could we expect to understand the important concept that "the chi-square statistic compares observed frequencies with expected frequencies" if we have no idea what expected frequencies are and how they are obtained?

There is another important reason why it is useful to know how to calculate expected frequencies: They can quickly tell us if we should be using the chi-square statistic at all. It is best to know this early in the data analysis process to avoid wasting time on a test that is inappropriate. As we noted earlier, the chi-square statistic is not appropriate for all analyses of the relationship between two nominal level variables. Its formula will not produce accurate results and it should, therefore, not be used if either of two situations exist:

1. When, in a 2-by-2 (four-cell) table, one or more cells has an expected frequency of less than 5.
2. When, in a table that is larger than 2-by-2, more than 20 percent of the cells have expected frequencies of less than 5.

A handy check on whether there is a problem with small expected frequencies in any cross-tabulation table can be performed by locating the cell with the smallest expected frequency. To do this, locate the row with the smallest marginal total and the column with the smallest marginal total. The cell with the smallest expected frequency is at the intersection of that row and column. The cell's expected frequency can be determined with the expected frequency formula. If an expected frequency in a 2-by-2 table is 5 or more, it is safe to use the chi-square statistic. If the expected frequency is under 5, chi-square should not be used.

In larger tables, additional "checks" will have to be made to see if the second "no-go" situation exists. If it does, it may still be possible to use chi-square by combining adjoining cells to form a table with fewer cells with sufficiently large expected frequencies. This process, called *collapsing,* should be done using common sense and logic and in a way that does not discard any more data than absolutely necessary. Values of a variable that are thus combined should possess a strong logical similarity to each other.

For example, in one research study that is attempting to learn if racist personnel actions have occurred within a certain corporation, it may be acceptable to collapse the original five values of the variable *race* into the two values *white* and *nonwhite* in order to have sufficiently large expected frequencies to be able to use the chi-square statistic. In another study, however, that is attempting to determine which race is least likely to enter a management position, collapsing the five values for the variable *race* would sacrifice data that are essential for hypothesis testing. Another statistical test should be used.

Meaningfulness of Chi-Square

Returning to our original example, which examined the relationship between type of treatment for alcoholism and treatment outcome, the corresponding p value for our chi-square value is very small (.005), which is what we had hoped to find. Since we did not specify a different rejection level, our cutoff point for rejection of the null hypothesis would be the customary .05, the point where we would conclude that chance would produce that large a difference between the observed and the expected frequencies less than five times out of 100. In fact, since .005 is much (ten times) smaller than .05, we would be very safe if we were to reject the null hypothesis that there is no relationship between the two variables. A p value of less than .005 means that chance would produce that large a difference between the observed and expected frequencies less than five times out of 1,000.

Can we now say that we have statistical support for our one-tailed research hypothesis? Not yet. Chi-square is just a statistical test; it has no way of knowing what we predicted in our one-tailed research hypothesis. It simply checks to see if the difference between the observed and expected frequencies is large enough that it probably is not the work of chance. In our example, a large chi-square value and a corresponding low p value would have occurred if either those clients in group treatment had a much better record of abstinence or if the better record had been among those clients in individual treatment. Statistical analysis told us that the variables probably are related, but it did not (and cannot) tell us the *direction* of that relationship. Was it in the direction that we hypothesized or in the opposite direction?

To find out if we have statistical support for our one-tailed hypothesis, we would compare the dummy table (Table 7.1) that we created to represent our one-tailed research hypothesis with Table 7.4. We would ask, "Where are the disproportionately large observed frequencies?" (A disproportionately large observed frequency is one that is larger than its expected frequency.) Are they in the cells where we predicted they would be (*a* and *d*) or are they in the opposite cells (*b* and *c*)?

The disproportionately large number of cases are in cells *a* and *d* in Table 7.4, just where we predicted they would be. Thus, we can reject the null hypothesis and claim support for our one-tailed research hypothesis. If the disproportionately large observed frequencies had occurred in cells *b* and *c* (where clients in individual treatment who abstained from alcohol and clients in group treatment who did not are displayed, respectively), we would have reason to believe that the variables are related but in the direction *opposite* to that which we predicted.

Having merely demonstrated a statistically significant association between the two variables, we are still a long way from saying that alcoholic clients should be put into group treatment because they are less likely to drink again. True, the difference between the percentages of abstinence in our two subsamples was pretty substantive (66.7 percent for group treatment versus 35.7 percent for individual treatment). However, before we can claim that we have uncovered a relationship that would hold for all or even any other alcohol treatment facilities, we want to be very sure that something unique did not exist within the population from which our sample was drawn (i.e., our agency).

For example, perhaps most of the alcohol and substance abuse workers in the agency were hired because they are especially adept at working with groups and they have less skill in individual counseling; or perhaps there was an unwritten policy that only those clients with the most severe alcohol problems are assigned to individual treatment. Either fact might explain the relatively high rate of success among clients who were in group treatment. We must remember that chi-square has demonstrated only an association between the two variables, not causation. Our statistical testing has only demonstrated that chance is an unlikely explanation of the apparent association between the two variables. It has done nothing to eliminate the other three alternative explanations (rival hypotheses, design bias, and sampling bias).

The most we can say about our finding of a statistically significant relationship between the two variables is that, most likely, the two variables really are associated, at least in the agency from which the sample was drawn. We cannot say with certainty, however, why this relationship occurred, nor can we predict if it will be found again in different agency treatment settings with different staff training, agency policies, and type of group or individual treatment used, for example. Maybe we are "on to something," but only replication and other research studies—perhaps using more powerful research designs and more powerful statistical analyses—will provide further evidence that the original one-tailed research hypothesis is, indeed, correct.

Meaningfulness and Sample Size.   As we suggested in Chapter 6, sample size influences whether or not a statistical significance will be achieved. When using any statistical test, the larger the sample size, the greater the chance of acquiring statistical support for the decision to reject the null hypothesis. Chi-square is no exception—large samples make it more powerful, sometimes too powerful. With a very large sample, it is unlikely that statistical significance will *not* be achieved, even if the actual percentage differences between the expected and observed frequencies are small.

When we see that a chi-square test has demonstrated the presence of a statistically significant relationship between two variables, it is easy to be impressed, but we need to look carefully at the sample size (N) that produced it. Like other statistical tests, a chi-square analysis can identify a statistically significant relationship between two variables, but it may be a statistically significant *weak* relationship.

A continuation of the example that we have been using throughout this chapter will make our discussion clearer. Let us suppose that we tried to replicate the results of our earlier research study. This time, a sample of 200 clients being treated for alcoholism were followed over a two-month period. The results of our second study are displayed in Table 7.5.

When a chi-square test is computed for the data in Table 7.5, p turns out to be greater than .10 (for a one-tailed hypothesis). Because the likelihood that chance might have produced the differences between the observed and expected frequency for these data is relatively large, we would lack sufficient statistical support to be able to reject the null hypothesis (.10 is greater than .05).

Since the evidence in support of our research hypothesis is now conflicting, we might want to attempt our study for a third time, this time with a very large sample. Let us suppose that we used data on not 200 clients, as is displayed in Table 7.5, but on 10 times more—2,000 clients! The data from this third study are displayed in Table 7.6.

TABLE 7.5 Observed frequencies and (percentages): Type of treatment by client outcome

Type of Treatment	Results		Totals
	Success	Failure	
Group	30 (60.0%)	20 (40.0%)	50 (100%)
Individual	80 (53.3%)	70 (46.7%)	150 (100%)
Totals	110	90	200

$\chi^2 = .429$, $df = 1$, $p > .10$ (direction predicted)

TABLE 7.6 Observed frequencies and (percentages): Type of treatment by client outcome

Type of Treatment	Results Success	Failure	Totals
Group	300 (60.0%)	200 (40.0%)	500 (100%)
Individual	800 (53.3%)	700 (46.7%)	1500 (100%)
Totals	1,100	900	2,000

$\chi^2 = 4.29$, $df = 1$, $p < .025$ (direction predicted)

A close look at Tables 7.5 and 7.6 reveals that the *percentages* of the 2,000 clients in the respective cells in Table 7.6 were exactly the same as those within the comparable cells in our sample of 200 cases in Table 7.5. In both hypothetical studies, 60 percent of clients in group treatment abstained from drinking compared with 53.3 percent of clients in individual treatment. Both 2-by-2 tables have one degree of freedom. But the differences in the chi-square values is very different. The value is ten times larger in the study using the larger sample than in the one using the smaller sample.

The *p* value also is very different (less than .025 with the sample of 2,000, compared with greater than .10 with the sample of 200). The relationship between the two variables was not statistically significant using the data in the study with 200 cases, but it was in the study with 2,000 cases. If we had used 20,000 clients, the chi-square value would be 42.9; if we had used 200,000 clients, it would be 429, and so on. Even though the percentages within the tables did not change, the strength of the association would be identical in each instance.

There is a logical reason why we *should* be more likely to be able to reject the null hypothesis when larger samples are used. When smaller samples are used (the "short run" mentioned in Chapter 5 when we discussed the role of probability in hypothesis testing), a difference of 6.7 percentage points (60.0 - 53.3 = 6.7) may have been the work of chance. With the much larger samples, the "law of averages" (the "long run") should have taken over; the differences in success for the two types of treatment should have disappeared if no real relationship between them and treatment success exists.

The fact that the 6.7 percentage difference still exists for the very large samples suggests that the relationship between the two variables is a real one, so we probably should reject the null hypothesis. However, we also should remember that the relationship between the variables is relatively weak, because it "shows up" only when very large samples are used.

Presentation of Chi-Square

The actual presentation of our findings using the chi-square statistic is straightforward. First, the cross-tabulation table is displayed that contains the observed frequencies. The chi-square statistic (χ^2), the degrees of freedom (*df*), and the probability of chance (*p*) associated with the χ^2 value are placed at the bottom of the table. These three pieces of information would then be printed at the bottom of the table as they are found in Box 7.1; that is:

$$\chi^2 = 8.24,\ df = 1,\ p < .005$$

THREE OR MORE VARIABLES

Frequently we wish to focus on the relationship between two variables, but we may be concerned that a third variable might "explain" an apparent relationship that we might be able to demonstrate statistically in some way. In the example we have been using, we have been seeking to document the existence of an association between the two variables *type of treatment* and *client success*. We may wonder if a third variable, for example, *a different level of motivation among clients in group treatment than among those in individual treatment,* may explain why clients in group treatment appear to do better than those in individual treatment. We would like to control for the effects of this third variable (called a *control variable*) to get a better picture of the true relationship between type of treatment and client outcome. There are two ways that this can be accomplished using the chi-square statistic.

Use of Two or More Tables

One way to explore the effect of a third variable is to divide our clients into all the categories of the third variable and examine the relationship between the two main variables, while controlling for the third one. Let us suppose that, in still another replication of our research study, data were collected and analyzed using 100 research participants. They were placed into a cross-tabulation table, and a chi-square analysis was performed, as displayed in Table 7.7.

We could divide our sample into the two subcategories of our third variable, high motivation for treatment and low motivation for treatment. We would then construct two separate tables (one for clients with high motivational levels

TABLE 7.7 Observed frequencies: Type of treatment by client outcome

Type of Treatment	Results		Totals
	Success	Failure	
Group	15	35	50
Individual	25	25	50
Totals	40	60	100

$\chi^2 = 3.376,\ df = 1,\ p < .05$ (direction predicted)

TABLE 7.8 Observed frequencies: Type of treatment by client outcome for highly motivated clients

| Type of Treatment | Results | | Totals |
	Success	Failure	
Group	7	18	25
Individual	13	12	25
Totals	20	30	50

$\chi^2 = 2.084$, $df = 1$, $p > .05$ (direction predicted)

and one for clients with low motivational levels) to look at the relationship between the two variables *type of treatment* and *client outcome*. The results might might look like those in Table 7.8 (high motivation for treatment) and Table 7.9 (low motivation for treatment).

The high probabilities of chance having produced the frequencies in Tables 7.8 and 7.9 reflects a very different finding from that in Table 7.7. Controlling for client motivation for treatment has caused our initial apparent relationship to nearly disappear. It is likely that the apparent relationship between the independent and dependent variables was not a real one, and that differences in the motivational levels of clients between the two types of treatment only made the independent and dependent variables appear to be related.

The original relationship will not always disappear when we control for a third variable; it may continue to exist in the same direction among all the values of that variable. If this happens, we would conclude that the third variable probably does not play a role in explaining the original relationship.

The relationship between the two original variables sometimes may appear stronger when a third variable is controlled. In such instances, the third variable is what we refer to as a *suppressor* variable (or an obscuring variable), one that causes us to underestimate the actual strength of the association between the independent and dependent variables.

When a third variable is introduced, a third result also may occur. The relationship between the independent and dependent variables may remain statistically

TABLE 7.9 Observed frequencies: Type of treatment by client outcome for low motivated clients

| Type of Treatment | Results | | Totals |
	Success	Failure	
Group	8	17	25
Individual	12	13	25
Totals	20	30	50

$\chi^2 = .750$, $df = 1$, $p > .10$ (direction predicted)

significant among the different categories of the control variable, but it may be significant in one direction with one or more categories of the variable and significant in the opposite direction with other categories. When this happens, it is usually not possible to summarize the findings easily; the relationship between the independent and dependent variables has to be described for each category of the control variable. The third variable is said to further specify the relationship between the first two variables, and it is therefore called a *specifying* variable.

Use of Multiple Cross-Tabulation Tables

There is a second method for exploring the effect of other variables on the relationship between the independent and dependent variables. It employs the use of a multiple cross-tabulation table and, unlike the first method, involves performing the data analysis using a single table. Multiple cross-tabulation involves spreading out the cases in the original cross-tabulation table among a larger number of cells. (This is the opposite of the process of collapsing that we described earlier.) Two (or more) variables are displayed along one axis or even along both axes. Table 7.10 is a multiple cross-tabulation table using the same data that were used in Tables 7.8 and 7.9.

Notice that the results of statistical analysis (*p* values) displayed in Table 7.10 are similar to those in Tables 7.8 and 7.9. The two methods for control of the third variable produced similar results; in both methods of analysis, we would be unable to reject the null hypothesis (using the .05 rejection level).

If we had performed our statistical analysis with pencil and paper, the cross-tabulation table (Table 7.10) could be studied to obtain a better understanding of how motivational level for treatment may influence the relationship between the other two variables. For example, we could look to see which cells seem to contribute the greatest amount to the chi-square value and in which ones the observed frequencies are larger or smaller than the expected ones.

TABLE 7.10 Observed and (expected) frequencies: Type of treatment and motivational level by client outcome

Type of Treatment	Motivational Level	Results		Totals
		Success	Failure	
Group	High	7 (10)	18 (15)	25
Group	Low	8 (10)	17 (15)	25
Individual	High	13 (10)	12 (15)	25
Individual	Low	12 (10)	13 (15)	25
Totals		40	60	100

$\chi^2 = 4.334$, *df* = 3, *p* > .05 (direction predicted)

Problems with Sizes of Expected Frequencies

It is not always possible to control for one or more variables using either two or more tables or multiple cross-tabulation. When data are spread out within larger cross-tabulation tables or within more of them, a situation may be produced in which the size requirements for expected frequencies for using chi-square cannot be met.

In Table 7.10, the 100 cases are distributed among eight cells. The expected frequencies of the right cells in Table 7.10 would thus be smaller than the expected frequencies of the four cells in Table 7.7. They remained sufficiently large to justify the use of chi-square, but only because the number of cases in group and individual treatment were equal and the overall number of successes and failures were similar. If the cases had not split relatively evenly into the various value categories, the expected frequencies may have been less than five in over 20 percent of the cells. Then we could not have used the chi-square test.

Because the use of two or more tables or the use of multiple cross-tabulation tables spreads out the cases into a larger number of cells, they can be used only when relatively large samples are available and cases break fairly evenly among values of the variables. Even then, if one or more variables examined has been measured in a way that results in a large number of different value categories, the values may have to be collapsed so that the cross-tabulation table has fewer cells (and sufficiently large expected frequencies) before we can use the chi-square test.

Only the expected frequency size requirement of chi-square prevents the more frequent use of multiple cross-tabulation tables for examining the effect of other variables on the relationship between the independent and dependent variables. Theoretically, several variables could be combined along the left of the table and several more could be combined along its top. To do this, however, a very large sample would be required, larger than is found in most social work research situations.

MICRO EXAMPLE

Background

Amelia is a social worker who works in a state in-patient hospital. Her main duty is to do readmission intake interviews with previous patients. From five years of experience, she has observed that a large number of patients who are readmitted to the hospital had been previously discharged to live with their relatives. Knowing that her social work colleagues doing discharge planning also made frequent use of boarding homes for discharged patients, she wondered why she was was not seeing more readmissions among those patients who were discharged to boarding homes. She wondered if there might not be a relationship between patients being readmitted to the hospital and the place to which they had been discharged (boarding home versus relatives).

Hypothesis

Amelia read all the literature on the topic that was available to her. Based on a general consensus among other social work practitioners, previous research findings, and her own subjective feelings and personal observations, she set out to design and implement a small-scale research study that would gather data to test her one-tailed hypothesis:

> Patients discharged to boarding homes have a lower rate of readmission to the hospital than patients discharged to live with their relatives.

Methodology

Amelia devised a simple research design to test her one-tailed hypothesis. She gained permission from her supervisors to select a 10 percent random sample of all the patient files from patients discharged during the last 18 months. Using a standardized data collection instrument that she constructed, she gathered data on a wide variety of demographic variables on the 148 patients (10 percent of 1,480 patients = 148 patients) who were discharged to boarding homes and the 250 patients (10 percent of 2,500 patients = 250 patients) who were discharged to relatives. Her total sample was 398 patients (148 + 250 = 398). The dependent variable in her hypothesis was the *admission status of the patient* (readmitted/not readmitted). The independent variable was the *patient's discharge status* (to boarding home/to relatives).

Findings

Table 7.11 presents Amelia's findings using the chi-square procedures presented in this chapter.

Conclusions

What did Amelia learn from testing her one-tailed hypothesis using the chi-square statistic? From her general knowledge of hypothesis testing, she knew that $p < .01$ was pretty impressive. It meant to her that the differences between the observed and expected frequencies probably were not the work of chance: The

TABLE 7.11 Readmission to hospital by discharge status

Discharge Status	Readmission?		Totals
	Yes	No	
Boarding Home	25	123	148
Relatives	71	179	250
Totals	96	302	398

$\chi^2 = 6.113$, $df = 1$, $p < .01$ (direction predicted)

likelihood that chance produced the apparent relationship between the two variables was less than one out of 100. She was able to reject the null hypothesis and
conclude that there was a statistically significant relationship between the two
variables. Consequently, she had statistical support for her one-tailed hypothesis.

Amelia also knew that chi-square analysis entails looking not only at the issue
of statistical significance but also at the question of whether the relationship
between the two variables was in the hypothesized direction. Chi-square analysis
is concerned primarily with the differences between the observed and expected
frequencies for all the cells. A difference is only a difference, whether it suggests
numbers that are smaller or larger than predicted. Either by looking for the cells
where disproportionately large observed frequencies occur or by examining percentages, Amelia knew that she must determine if the association was in the
predicted direction.

Using Table 7.11, Amelia was able to determine that roughly 17 percent (25 of
148) of patients discharged to boarding homes had been readmitted to the hospital,
compared with 28 percent (71 of 250) of those discharged to relatives. These
two percentages, 17 and 28, were consistent with the direction of her hypothesis; patients discharged to boarding homes were less likely to be readmitted than
patients discharged to relatives.

Before Amelia drew any conclusions about the "meaning" of the statistical
significance between the two variables, she knew that she must acknowledge the
limitations of her research design in interpreting her findings. She had used a
standardized and structured data collection instrument. However, the validity and
reliability of the data in the patient records might be a problem; in addition, there
might be other factors relating to bias.

Because of the lack of an experimental design, there was also a long list of
other variables that might have affected readmission of the patients in her study.
They included patient diagnosis, length of first hospitalization, availability of after-
care services, patient's use of medication, and myriad other factors that she had
no reason to believe were equally represented in the two groups (boarding home/
relatives) of patients. The possible confounding effects of the interaction of several
of these other variables on readmission were even more overwhelming for Amelia
to consider.

What did her findings really tell her about the one-tailed hypothesis? The goal
of chi-square is to acquire evidence for or against the existence of an association between two variables. Cause and effect knowledge was not a possibility
from the beginning. This was due partly to the absence of an experimental design
and partly to the limits of the chi-square analysis itself. What Amelia learned was
that, for whatever reasons, patients discharged to boarding homes from her particular hospital were not as likely to be readmitted as those patients who were discharged to relatives.

Amelia did not limit her analysis to only the relationship between the independent and dependent variables. She also gathered data on patient diagnosis and
length of first hospitalization. She could examine the relationship between these
other variables and the dependent variable as well, using more complex analyses

or other, more appropriate statistical tests. The patient's records might have yielded insights into other variables that went into the decision to live with relatives or in a boarding home; these data could be used to enhance the results of her analysis and to shed more light on the statistical findings.

Implications

Given all the limitations that tempered the interpretation of Amelia's findings, it might seem that, on the surface, they were of little value. The likelihood of the existence of many other variables that might have affected the dependent variable would seem to discount even further the utility of her research findings for social work practice. Yet, despite a less than perfect research design and the use of a relatively low-powered data analysis, several possibly valuable practice implications emerged.

The contributions of Amelia's findings to our professional body of knowledge would be quite limited. Conditions relating to admission, treatment, discharge, and readmission at Amelia's hospital are likely to be different from those in other in-patient psychiatric settings. It might be possible to generalize her findings beyond her particular hospital, but only with a great deal of caution.

Most of the benefits of Amelia's research study would be felt on a local level. One of the major benefits of identifying a relationship between two variables is that it can improve our ability to predict the future. Knowing (even without fully understanding *why*) that patients discharged to boarding homes in Amelia's hospital are less likely to be readmitted than patients discharged to relatives could be valuable to her social work colleagues doing discharge planning. Patients discharged to live with relatives could be perceived as being at higher risk of readmission.

A few implications could be of immediate use to Amelia. Based on her findings in Table 7.11, she began to ask herself these questions:

1. Should I try to find boarding home placements for more of my clients?
2. Should I endeavor as a professional social worker to work toward the creation of more boarding home facilities?
3. Should I endeavor to provide additional sources of support for patients going to live with relatives in order to reduce the likelihood of their being readmitted?

These applications of the research findings are still only questions. None of them indicates that social work practitioners should be prepared to make drastic modifications in their service delivery methods without careful consideration and, what is more important, without evidence of further research findings.

Many other questions might emerge from the data presented in Table 7.11. Some may relate to treatment issues, whereas others would have administrative implications.

Amelia should use great caution in interpreting her research findings. Any major changes in the delivery of social work services may have a negative effect on the lives of patients, and she should remember that implementations derived from chi-square findings can be risky. Chi-square analyses are sometimes best used to tentatively identify associations between variables and to formulate critical questions that can subsequently be answered through more high-powered statistical analysis.

MACRO EXAMPLE

Background

Juan is a social worker employed by a legislative committee. He is attempting to assist in the passage of an increase in the state sales tax, which will provide additional revenues for public schools. As part of his job, he recently began gathering general personal demographic data on state legislators in order to gain insight as to why they might vote for or against the proposed tax bill.

After he had examined some available data and hearsay on about 30 legislators, it seemed to him that legislators who favored the bill tended to have children who currently attended public schools; those legislators who were on record as not supporting the bill tended not to have children currently in public schools. Juan wondered whether he could find statistical support for a relationship between the dependent variable (*support or nonsupport for the bill*) and the independent variable (*use or nonuse of public schools*).

Hypothesis

After discussion with committee members and a review of the literature available on legislators' voting patterns on various tax issues, Juan concluded that he could justify formulation of a one-tailed hypothesis:

> Legislators who currently have children attending public schools will
> more likely support the bill than will legislators who do not have children
> currently attending public schools.

Methodology

Juan continued to gather data on the state's legislators, but he made a special effort to learn about and systematically record data on his independent and dependent variables. He identified 160 legislators who had publicly stated support for or opposition to the tax bill. Of these, he obtained sufficient current biographical data on 125 (78 percent) of them and was able to conclude with reasonable certainty whether or not they were currently sending their children to public schools.

Findings

Table 7.12 presents Juan's findings using the chi-square statistic as presented in this chapter.

TABLE 7.12 Legislators' support for tax bill by whether they use public schools

Use Public Schools?	Support for Bill?		Totals
	Yes	No	
Yes	39	36	75
No	21	29	50
Totals	60	65	125

$\chi^2 = .834$, $df = 1$, $p > .10$ (direction predicted)

Conclusions

Juan was somewhat dismayed at the results of his chi-square analysis, yet he was grateful that he had not trusted his subjective hunches before he spoke prematurely to the legislative committee. Based on his analysis, he would have had a fairly high likelihood of committing a Type I error if he had rejected the null hypothesis and claimed the existence of a relationship between the independent and dependent variables.

With one degree of freedom and a one-tailed hypothesis, he knew his chi-square value should have been at least 2.71 (Appendix B) to reject the null hypothesis at the customary .05 level of statistical significance. He now had reason to believe that the legislators' use or nonuse of public schools for their children was probably not associated with their voting preference on the tax bill.

Other independent variables may have reflected a stronger degree of association with the dependent variable—the legislators' voting record on all tax bills, their perceptions of the fairness of a sales tax versus other sources of revenue, marriage to a spouse who was teaching in the public school system, and so on. Juan could be reasonably certain, however, that—based on his study—there was no reason to conclude that legislators' support for or opposition to the tax bill was related to whether or not they had children currently in the public school system.

Implications

The lack of statistical support for Juan's one-tailed hypothesis in no way negated the value of his research study. His small-scale study, while limited and singularly focused, was useful. He avoided making an erroneous conclusion based on inadequate evidence that might have resulted in an inappropriate lobbying strategy. Portraying the motivation of legislators who opposed the bill as "self-serving" would have been ill advised. Juan wondered whether it would be productive to undertake further statistical analyses using the legislators' other demographic characteristics as independent variables. He could now spend his time more efficiently by pursuing other avenues to gain insight into the legislators' positions on the proposed tax.

CONCLUDING THOUGHTS

The fact that chi-square requires only nominal level data makes it ideally suited to many of the hypothesis testing needs of social work researchers and practitioners. Unfortunately, there are certain errors commonly made in its use. The first two, suggested earlier in this chapter, involve making "too much" of a finding of statistical significance by either (a) implying that the relationship is one of more than just association or (b) overestimating the importance of findings that reflect weak relationships between variables where a finding of significance was virtually inevitable because of the use of very large samples.

A third common error is made when chi-square is used when one or both variables is more than at the nominal level of measurement. To use chi-square in situations where ordinal, interval, or ratio measurement is available is a waste of measurement precision. While a chi-square value can be computed, the test will treat the different value categories as if they are only differences in kind, and it will ignore the fact that they reflect differences in the quantity of the variable. Unless both variables are only at the nominal level of measurement, there is very likely to be a better (often a more powerful) test that can and should be used.

STUDY QUESTIONS

1. Why is the frequent use of chi-square in social work research both good and bad? Explain.
2. How is the chi-square test particularly well suited to social work research?
3. What do the numbers in each of the cells in a cross-tabulation table mean?
4. What is a dummy table, and what does it reflect?
5. Can chi-square tell us whether one variable causes variations in the second variable? Explain.
6. What are expected frequencies, and how are they used in chi-square testing?
7. How do degrees of freedom affect whether a chi-square value of a given size (e.g., 10.00) will be considered statistically significant?
8. What are the minimum expected frequency requirements for the use of chi-square?
9. What is the final step in the process for determining if chi-square has demonstrated support for a one-tailed hypothesis? Provide examples.
10. What are two ways that chi-square can be used to examine the relationship between two variables while controlling for the effect of a third variable? Provide a social work example of when this might be desirable.

chapter **8**

Correlation

Chapter 7 presented a method to analyze the relationship between two nominal level variables using a nonparametric statistic known as chi-square. This chapter examines a method to analyze the strength and direction of a linear relationship between two interval or ratio level variables using a parametric statistical test known as Pearson's r. It is helpful to visualize the way that r works through the use of scattergrams.

THE SCATTERGRAM

Unlike the other graphs discussed in Chapter 2, which displayed how many times a value occurred for a given variable, *scattergrams* display all individual case values. They also portray the case values for two variables simultaneously.

Each dot on a scattergram represents the intersection of two values (one for each variable). For example, if 23 clients completed a self-administered standardized measurement instrument that measures self-esteem and also completed a standardized self-administered measurement instrument that measures client satisfaction with social work services, a scattergram could portray each person's score on both variables (self-esteem and satisfaction with services) with a single dot. The scattergram would have 23 dots—not 46—to represent each of the 23 cases and their combination of measurements for the two variables.

Uses of Scattergrams

A scattergram has three general uses. First, it can be used to portray a possible relationship between two variables. Second, it can be used to portray graphically a relationship that has been demonstrated to exist through the use of statistical analyses. It also can be used to identify and to present patterns of a possible relationship between two variables that may exist within certain subsets of a larger data set but that tend to get "lost" when the total data set is statistically analyzed. For example, there may be no apparent relationship between age (first variable) and job satisfaction (second variable) among blue-collar workers when we statistically analyze the data from a large research sample of workers between the ages of 18 and 70. When the same data are displayed using a scattergram, however, a relationship may be quite evident within the range of 36 to 45 years of age.

Drawing Scattergrams

Scattergrams use two perpendicular lines (*x*-axis and *y*-axis), as illustrated in Figure 2.1 in Chapter 2 and Figure 8.1. However, unlike most other graphs, scattergrams plot the values of the first variable along the *x*-axis and the values of the second variable along the *y*-axis. In scattergrams designed to detect and/or display a possible relationship between two variables, the first variable is customarily identified as the *predictor* variable and the second as the *criterion* variable. These terms are used, rather than the terms *independent* and *dependent,* respectively, that we used in Chapter 7, because they more accurately reflect the nature of the relationship between variables that is examined using scattergrams and correlation.

The *x*-axis is used to plot the values of the predictor variable and the *y*-axis is used to plot the values of the criterion variable, as shown in Figure 8.1. If neither variable is clearly the predictor variable or the criterion variable, either variable can be plotted along either axis.

The values for the predictor and criterion variables for a particular graph may or may not have identical intervals. This presents no special problem. The

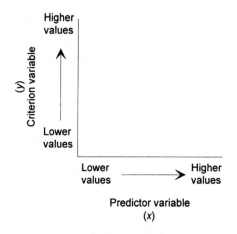

FIGURE 8.1 Basic outline of a graph for depicting values for two variables

values for the predictor variable may be marked off at intervals of five units, for example, while the values for the criterion variable may be marked off at intervals of ten units. Most statistical software packages that produce scattergrams automatically draw them to scale based upon the raw data that have been entered.

EXAMINING RELATIONSHIPS USING SCATTERGRAMS

Suppose that we are interested in studying the relationship between the two variables *number of treatment sessions a client attends* (predictor variable) and his or her *self-esteem level* (criterion variable). We might optimistically hypothesize that, among clients being treated for low self-esteem, there would be a positive association between current self-esteem levels and number of treatment sessions attended; that is, the more treatment sessions clients attended, the higher their self-esteem, and vice versa. The number of treatment sessions attended (the predictor variable) and the scores each client achieved on the measuring instrument that measures self-esteem (the criterion variable) could then be displayed, as in Figure 8.2.

Each dot represents an individual client's level on the self-esteem measuring instrument (*y*-axis) and the number of treatment sessions attended (*x*-axis). As can be seen easily in Figure 8.2, 10 individuals (dots on the graph) represent our small treatment sample. Notice that Lydia, who attended 10 treatment sessions, had a self-esteem level of 30, but William, who attended 18 sessions, had a self-esteem level of 40. The fact that the dots in Figure 8.2 are scattered somewhat (rather than falling in a straight line) suggests that the relationship between the two variables within our sample was far from perfect.

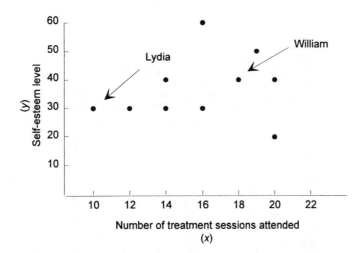

FIGURE 8.2 Scattergram: Number of treatment sessions attended and self-esteem level

Perfect Relationships

An example of a perfect relationship between a predictor variable and a criterion variable can be seen in Table 8.1. The table contains two variables, *clients' motivation for treatment* and *clients' level of functioning,* and it summarizes hypothetical data for a sample of ten clients. For every client value of the predictor variable (client motivational level for treatment), there is a corresponding client value of the criterion variable (client functioning level).

Note in Table 8.1 that if, prior to data collection, a one-tailed research hypothesis such as "Clients with higher motivational levels for treatment will have higher functioning levels than clients with lower motivational levels" had been formulated, the data would lend support to that hypothesis. A relationship between the two variables in Table 8.1 is evident because, *without exception,* higher motivational levels for treatment are associated with higher levels of functioning, and vice versa. Floyd, for example, scored lowest on both motivational level for treatment (1) and functioning level (2). Jane scored second lowest on both levels (scores of 2 and 3, respectively), and Lynne scored highest on both (scores of 10 and 11, respectively). This perfect relationship also can be depicted by means of a scattergram, such as the one illustrated in Figure 8.3.

In Figure 8.3 the horizontal axis represents the clients' individual scores, or measurements, of their motivational levels for treatment (x-axis), while the vertical axis represents the individual scores of their functioning levels (y-axis). Each dot represents a pair of scores for one person. Thus, in Figure 8.3, there are 10 dots representing the 10 clients and their respective pairs of scores on the two variables. The dots, if connected, would form a straight line, indicating that the two variables are perfectly correlated. Such perfection is almost never seen in data drawn from social work research studies. It is used here to illustrate the concept of correlation in its most vivid form, prior to our discussion of the kind

TABLE 8.1 Motivation for treatment and functioning levels ($N = 10$)

Name	Motivational Level (X)	Functioning Level (Y)
Floyd	1	2
Jane	2	3
Robert	3	4
Sue	4	5
Herb	5	6
Bill	6	7
Margareta	7	8
Ann	8	9
Dorothy	9	10
Lynne	10	11

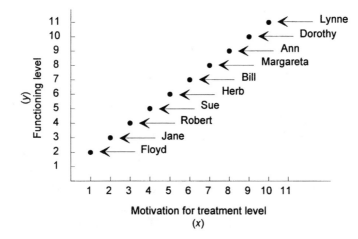

FIGURE 8.3 Scattergram of perfect positive relationship between two variables: Motivation for treatment and functioning levels (from Table 8.1)

of correlations that we more commonly find in social work research and practice situations—nonperfect relationships.

When perfect correlations do occur in social work research, we must always wonder whether they really reflect a meaningful relationship between two variables or whether what were assumed to be two variables were really just two different kinds of measurements of the same variable. For example, in some social work agencies, we probably could demonstrate a nearly perfect correlation between workers' sick leave days earned and total number of clients seen by them, but such a correlation would be neither meaningful nor surprising. The "two variables" may be just two different measurements of the same variable—time on the job.

Strength and Direction. Figure 8.3 demonstrates two important characteristics of the perfect relationship between the two variables: strength and direction. With regard to *strength,* the overall relationship between the two variables could not be stronger, as all the clients' paired scores fall along a straight line. While Figure 8.3 is an example of a perfect positive relationship, Figure 8.4 is an example of a perfect negative relationship. In both figures, the dots fall in a straight line. In those more common situations where the relationship between variables is less than perfect, a line called a *regression line* (to be discussed shortly) can be drawn to represent the best fit among a scatter of dots.

In regard to the second characteristic of a correlation—*direction*—the relationship between the clients' motivation for treatment levels and their functioning levels, as displayed in Figure 8.3, can be described as positive. High values of *x* are associated with high values of *y,* and low values of *x* are associated

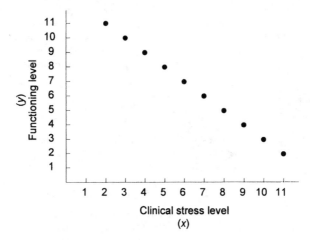

FIGURE 8.4 Scattergram of perfect negative relationship between two variables: Clinical stress and functioning levels

with low values of *y*. In a negative relationship, such as the one displayed in Figure 8.4, high values of one variable are associated with low values of the second. Figure 8.4 suggests that higher clinical stress levels (predictor variable) were associated with lower functioning levels (criterion variable).

Nonperfect Relationships

In social work research, some relationships may have no discernible direction as well as no strength; that is, there is no apparent relationship at all between the two variables. Figure 8.5 displays hypothetical data in which there is no perceivable relationship between the variables *self-esteem* and *aggression*. In other words, we cannot say that the more (or less) self-esteem that our clients have the more (or less) aggressive they are.

Most relationships reflect some degree of correlation, ranging from barely discernible to nearly perfect (nonperfect correlations). Figure 8.6 is a scattergram illustrating a nonperfect positive relationship between the variables *number of treatment sessions attended* and *client attitudes toward their peers*. The relationship is still positive, but it is not perfect, like the one illustrated in Figure 8.3.

Figure 8.6 shows that three clients were seen only once by a social worker, but one client (Sue) scored a 2 on the attitudes toward peers scale, another (Robert) scored a 4, and a third client scored a 6. So, the relationship between the variables is far from perfect, even though the dots fall in an overall positive direction (bottom left to upper right). In contrast, Figure 8.7 reflects no relationship at all between the variables *depression level* and *motivation for treatment*.

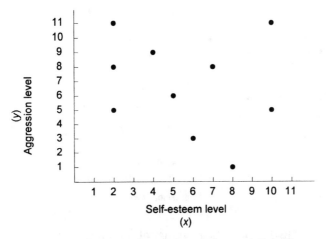

FIGURE 8.5 Scattergram of no relationship between two variables: Self-esteem and aggression levels

The dots in scattergrams (one value for the predictor variable *and* one value for the criterion variable for each case) can fall in a variety of patterns, such as a straight line, a "U" shape, and a "J" shape. Observing the pattern in which they fall can be useful for understanding and drawing conclusions about relationships between two variables.

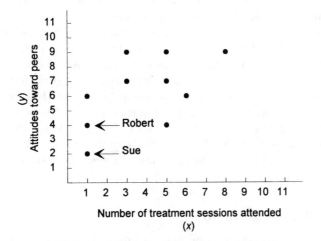

FIGURE 8.6 Scattergram of nonperfect positive relationship between two variables: Number of treatment sessions attended and attitudes toward peers

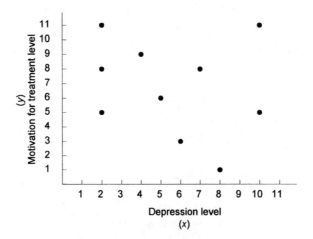

FIGURE 8.7 Scattergram of no relationship between two variables: Motivation for treatment and depression levels

INTERPRETING CORRELATIONS

Scattergrams are a somewhat crude way of displaying the paired values for two variables within a large data set. A far more efficient and accurate way of displaying the relationship between two interval or ratio level variables is through the use of a *correlation coefficient*. A correlation coefficient, expressed as the italicized letter *r,* provides a numerical indicator of both the strength and the direction of the relationship between two variables.

It is important to remember that *r* is the degree to which two variables are *linear;* that is, it is the degree to which a scattergram of a distribution of two variables approximates a straight line (perfect correlation). Of course, two variables may be correlated, that is, they may be related, but their relationship may not be a linear one. For example, the correlation (as reflected by connecting the dots in a scattergram) might be a series of waves (*cyclical correlation*) or a line that is curved like either an arch or an upside down arch (*curvilinear correlations*). The clear *pattern* of the dots and the fact that they form a line indicate the presence of a relationship between the variables, but it is not a linear (i.e., a straight line) correlation.

The Correlation Continuum

As Figure 8.8 shows, correlation coefficients range along a continuum from −1.0 (perfect negative) at one extreme to 1.0 (perfect positive) at the other extreme, with 0.0 (no linear correlation) at the midpoint. A correlation coefficient cannot be greater than 1.0 or less than −1.0.

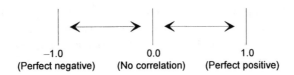

−1.0   0.0   1.0
(Perfect negative)   (No correlation)   (Perfect positive)

FIGURE 8.8 Correlation coefficient
continuum

The closer the numerical value of the correlation coefficient is to either extreme (1.0 or −1.0), the stronger the linear relationship between the two variables. For example, a coefficient of .92 is closer to a perfect correlation than coefficients of either −.60 or .60 and therefore suggests a stronger correlation than either of the other two. The closer the coefficient is to the middle of the continuum (0), the weaker the relationship between the two variables. A correlation coefficient that is close to 0 suggests that the linear relationship, if any, is quite weak.

The plus or minus sign of r indicates the direction of the relationship. For example, the correlation coefficient (r) between a social worker's skill level and years of professional social work experience might be $r = .85$, a positive relationship. The correlation between skill level and level of apathy regarding one's work might be $r = −.75$, indicating a fairly strong negative relationship; that is, persons who are most apathetic and uninvolved in their work are likely to be among those who are the least skillful, and vice versa.

What Is a Correlation Coefficient?

As we have suggested, the word *correlation* is a generic term used to describe the strength and direction of a linear relationship between two variables. Correlations are observed in different ways. Similarly, a correlation coefficient (ranging from $r = +1.00$ to $r = −1.00$) can be computed using different statistical formulas. The most commonly used one, which results in a Pearson's Product Moment Correlation Coefficient, or simply Pearson's r, requires that both the predictor and criterion variables be at the interval or ratio levels of measurement.

The Coefficient of Determination

The correlation coefficient produced by the formula for Pearson's r tells simply to what degree and in what direction two variables covary. But if the correlation coefficient (r) thus produced is squared (r^2), we can learn much more. The new number (r^2) indicates the proportion of the variation in one variable that can be "explained" by, or is related to, its relationship to the other variable. r^2 is called the *coefficient of determination*.

For example, an r value of .80 relating the number of treatment interviews and self-esteem levels means that 64 percent ($.80^2 = 64\%$) of the variation in self-esteem levels might be explained by the number of treatment interviews. The other

36 percent of variation in self-esteem levels (1.00 − 64% = 36%) remains unaccounted for. Our ability to estimate clients' self-esteem scores if we know the number of interviews they have had would be quite good, but certainly not perfect.

Correlation Is Not Causation

As should be evident by now, a correlation coefficient communicates the degree to which two variables covary in a linear fashion within a data set. What it cannot do is tell us that the value of one (criterion) variable has resulted *because* of the value of the second (predictor) variable. As we suggested earlier, correlation should not be used to make any statement about causation. Why not?

Unlike a research experiment in which the researcher introduces the independent variable, correlation analysis generally is used to examine patterns of relationships between variables for which data (on both the predictor and criterion variables) have already been collected. The researcher does not introduce or control measurements of the predictor variable; rather, its natural variation is observed along with that of the criterion variable. Thus any interpretation of correlation that implies causation goes beyond the capabilities of the statistic. Usually correlation is used to determine if any relationship exists (*descriptive research*), but it also sometimes is used to test theoretical assumptions about the nature of a relationship (*explanatory research*).

What is the relationship between correlation and causation? Correlation is necessary but not sufficient for proof of causation. This means that where causation exists between two variables, those same two variables always will be correlated. However, just because two variables may covary, there is no reason to believe that they possess a relationship that involves causation. There may be many other possible explanations for a correlation between them besides a cause-effect relationship.

Many pairs of variables tend to covary; they could be demonstrated to be correlated. However, any statement suggesting causation would be ludicrous. Examples include: (a) ice-cream sales as a predictor variable and drowning rates as a criterion variable, (b) bible sales as a predictor variable and whiskey sales as a criterion variable, and (c) diet drink consumption as a predictor variable and weight of individuals as a criterion variable. More often than not, a correlation between a predictor and criterion variable is easily explained by the effects of a host of other predictor variables. In the previous three examples, other predictor variables such as temperature, per capita income, and the greater tendency of overweight persons to watch their caloric intake, respectively, would help to explain the linear correlations.

Sometimes the influence of other predictor variables on a correlation is less obvious. We may have heard media exhortations to young people to complete high school and, if possible, to go to college. The appeal usually includes correlational data about the relationship between the variables *educational level* (predictor variable) and *lifetime earnings* (criterion variable). A statement is made that the farther people go in school, the more money they will make over their lifetimes. A cause-effect interpretation of correlation data is implied. The correlation between educational level and lifetime earnings may be undeniable. However, can it be interpreted to mean that going on in school will automatically increase one's earning potential? No!

The criterion variable *financial success in life* may be strongly correlated with other predictor variables such as *motivation, intelligence,* and the *socioeconomic status of parents.* All these variables could reflect a stronger correlation with lifetime earnings than would educational level. They may contribute to high levels of education *and* to lifetime earnings. The complex interaction of variables that relate to lifetime earnings argues against any statement implying a finding of causation, despite the fact that many different correlations could be demonstrated to exist. All other things being equal (and when does this ever occur?), the high school diploma or college degree might give one an edge in being financially successful. However, this is not a legitimate conclusion drawn from a simple correlation between the variables *educational level* and *lifetime earnings.*

A correlation coefficient is a concise summary of the strength and direction of the association between two interval or ratio level variables. It is a mathematical alternative to a scattergram for communicating to the reader of a research report how the values of one variable tend to be associated with the values of another variable. For example, it might be useful for the reader to know that, within a sample of employees studied within an organization, there was a high negative correlation between the variables *age* and *claims for disability insurance,* even though this phenomenon may not be typical or representative of social work agencies in general. Knowing this would help readers to better understand the unique research sample and setting of the study as well as to recognize differences between the context of the study's situation and their own.

COMPUTATION OF A CORRELATION COEFFICIENT

As with the chi-square statistic presented in Chapter 7, the *r* statistic can easily be computed with the aid of a computer. Box 8.1 is provided for readers who wish to see the mathematical foundations of the *r* formula. Box 8.1 uses ratio level data on the variables *unemployment rate* and *number of civil disturbances* for a given time period within five cities. The raw data are listed in the *x* and *y* columns in Table 8.2.

Relating *r* to Chance

How large an *r* do we need in order to conclude that we can safely reject the null hypothesis and be able to claim statistical support for the position that two variables may be related? As is true with other bivariate relationships, the correlation coefficient is judged to be sufficiently strong if it reaches or surpasses our predetermined level of statistical significance. Unless previously stated and

---

BOX 8.1
The Correlation Coefficient

Hypothesis: There is a positive association between the unemployment rate and the number of civil disturbances.

Null Hypothesis: There is no association ($r = 0$) between the unemployment rate and the number of civil disturbances.

Predictor Variable: Unemployment rate—ratio level.

Criterion Variable: Number of civil disturbances—ratio level.

Observed Frequencies: See Table 8.2.

Correlation Coefficient Formula:

$$r = \frac{N\Sigma XY - (\Sigma X)(\Sigma Y)}{\sqrt{[N\Sigma X^2 - (\Sigma X)^2][N\Sigma Y^2 - (\Sigma Y)^2]}}$$

Where:
r = Correlation coefficient
N = Number of cases
ΣXY = Sum of xy column
ΣX = Sum of x column
ΣY = Sum of y column
ΣX^2 = Sum of x^2 column
ΣY^2 = Sum of y^2 column

Substituting values for letters:

$$r = \frac{(5)(985) - (76)(53)}{\sqrt{[(5)(1290) - (76)^2][(5)(919) - (53)^2]}}$$

$$= \frac{897}{\sqrt{(674)(1786)}}$$

$$= \frac{897}{\sqrt{1203764}}$$

$$= \frac{897}{1097.16}$$

$$= .82 \text{ (correlation coefficient)}$$

Presentation of Results:

$$r = .82, \, p < .05$$

Conclusions: The null hypothesis is rejected and the one-tailed research hypothesis is supported. In short, there is a statistically significant association between the variables unemployment rate and the number of civil disturbances. The coefficient of determination, r^2, is $(.82)^2$ or .67. This indicates that 67 percent of the variance in civil disturbances can be explained by the unemployment rate.

---

justified, the .05 rejection level is used as the reference point for determining whether we can reject the null hypothesis.

The table of critical values of r (Appendix C) illustrates the strength of correlation coefficient that is required to achieve statistical significance at various rejection levels. For example, as Appendix C on page 217 shows, with a sample of 11

TABLE 8.2 Unemployment rate and the number of civil disturbances in five large cities

City	Unemployment Rate (X)	Civil Disturbances (Y)	X^2	Y^2	XY
Dallas	22	25	484	625	550
Chicago	20	13	400	169	260
Los Angeles	10	10	100	100	100
New York	15	5	225	25	75
Miami	9	0	81	0	0
	76	53	1290	919	985

individuals, a correlation coefficient of .6021 is required with a two-tailed hypothesis to reach statistical significance at the .05 rejection level (.7348 at the .01 level) and thereby to permit rejection of the null hypothesis. With a sample size of 102 cases, however, rejection of the null hypothesis is possible with a much weaker correlation coefficient (.1946 at the .05 level and .2540 at the .01 level). How can this be?

As with chi-square analyses, the likelihood of demonstrating statistical significance with r is related directly to sample size. It is more likely that chance will cause two variables to appear to be related with a small sample than with a larger one. With larger samples, an apparent relationship, even one that on the surface appears quite weak, is far less likely to be the work of chance. Tables such as Appendix C take this phenomenon into consideration—namely, that the correlation coefficients required to achieve various rejection levels vary with sample size.

Meaningfulness of r

In interpreting the strength of a correlation coefficient, we must also take into consideration the way in which a correlational analysis is to be used. Even when a statistically significant correlation coefficient is obtained, the r itself is not necessarily meaningful and may not represent a meaningful finding. A statistically significant, relatively low correlation coefficient may represent a meaningful finding in one study, but a relatively high correlation coefficient in another study may be relatively unimportant. Even identifying no association ($r = 0$) between two variables that are *believed* to be related may represent the most important finding of a study.

In interpreting a correlation coefficient, it is important not to treat it as if it were equivalent to interval or ratio level data or to make statements that in any way give this impression. For example, a correlation coefficient of .80 is not twice as strong as one that is .40. In fact, the .80 describes an association four times as strong ($.80^2 = .64$; $.40^2 = .16$; $.64/.16 = 4$) in its ability to account for the amount of variation in the criterion variable from the variation in the predictor variable.

It should also be remembered that in a correlation coefficient as strong as .80, there will be few exceptions to the pattern of relationships between values of one variable and values of the second variable; that is, virtually all high values of the first variable will be found in cases with high values of the second variable, and vice versa. A weaker correlation coefficient (such as .40) will have a much higher percentage of cases that are opposite to the overall direction of the association.

Presentation of r

When r is reported in a research report to describe the strength and direction between two interval or ratio level variables, a short narrative description is usually included to elaborate on its meaning. For example, the text might read: "Among the ten clients participating in the research study, the correlation between the number of treatment interviews they completed and their self-esteem scores was $r = .72, p < .01$ (one-tailed test). This means that their self-esteem was positively correlated with the number of times they were seen by the social worker." Or: "The correlation between the predictor and criterion variables for the ten clients was $r = .72, p < .01$, which indicates the presence of a statistically significant positive correlation between the two variables."

If more than just a few bivariate correlation coefficients (those reflecting the correlation between two variables) need to be reported, reporting them in tabular form may be helpful to the reader of a research report. A table designed for this purpose is called a *correlation matrix*. It lists all variables in a column on the left side and repeats them in a row along the top. The reader can find the direction and strength of a correlation between any two variables by noting the correlation coefficient that appears where the row in which the first variable appears intersects with the column headed by the second variable.

THREE OR MORE VARIABLES

Bivariate relationships frequently need further explication. Moreover, they can also be misleading. For example, we are unlikely to explain how long a client remains in treatment based solely on data on the severity of the client's presenting problem, the client's motivation for treatment, or any other single variable. Both a systems perspective and other theories of multiple causation argue that many variables work together to affect a single variable such as a human behavior or attitude.

The accuracy of explanation can be improved by expanding the pool of available data to include more than two variables at one time in the analysis. Decisions regarding the direction that this expansion should take as well as the additional sources and types of data needed to improve our explanative abilities are frequently the next critical steps along the data analysis continuum.

There are some hypothesis-testing situations where we conclude that there is one possibly intervening variable that cannot be ignored and that we have been unable to control for using our research design. For example, in testing for a possible correlation between the variables *age* and *amount of charitable contributions* to a family service agency, it may be necessary to use some method to control for the variable *per capita income.*

A relatively simple statistical test, *partial r,* is available for this type of situation. Data on all three variables can be entered into the *partial r* formula, and a correlation coefficient can be obtained between any two of the variables while the influence of the third is mathematically "controlled." Using *partial r* we would be able to determine how large and/or statistically significant a correlation between the variables *age* and *charitable contributions* would exist if the variable *per capita income* were *not* a factor in their relationship. A useful by-product of this analysis would be a correlation coefficient reflecting the correlation between the variables *per capita income* and *charitable contributions.*

A more sophisticated type of statistical analysis is available for situations in which we are interested in knowing the amount of variation of the criterion variable that can be explained by several predictor variables working in combination. *Multiple correlation* (to be discussed in greater detail in Chapter 12) looks at a whole set of predictor variables together in order to determine the degree that their *combined* variations correlate with different values of the criterion variable. It can be used to help us identify the best set of predictor variables for predicting values of the criterion variable. For example, it might tell us which three interval or ratio demographic variables might reflect the highest correlation with rate of homelessness, incidence of spouse abuse, or some other social problem. An "at-risk" population might thus be identified.

Multivariate (more than two variables) correlation analyses such as *partial r* or *multiple r* often are superior to bivariate analyses in hypothesis testing. They can easily examine the complexity of the interaction of three or more variables in a way that bivariate analyses cannot. They can also help avoid committing Type I errors by stumbling onto spurious (false) "relationships" that result simply from trying enough combinations of bivariate analyses until one finally reflects a statistically significant correlation. (We will return to this potential problem in Chapter 10.) Let us now see how bivariate correlational analyses can be used in social work practice situations.

MICRO EXAMPLE

Background

Leon is a social worker in a family service agency. He leads several treatment groups consisting of female adolescents. He recently became aware of the wide variation in verbal participation among the group members. While virtually all the members responded when spoken to, a few never made any unsolicited comments. He perceived that they had a low desire to get involved with the group. Over a period of several weeks, Leon made it a point to ask some of the nonverbal members why they rarely participated verbally in the group sessions. Of the seven members he asked, five replied with essentially the same answer: Each was an only child in her family and had been taught by her parents that it was not her role to initiate communication. Leon then asked three of the most verbal adolescents, who tended to dominate group discussions, how many siblings they had. Their responses were six, seven, and nine, respectively.

Based on his limited inquiries, Leon began to speculate on a possible relationship between a criterion variable, *number of unsolicited comments in group treatment* and a predictor variable *number of siblings in the family.*

Hypothesis

Leon conducted a quick literature review. He learned all that he could about such phenomena as social traits of only children, communication patterns among siblings, and variations in verbal participation in adolescent groups. Most of the literature seemed to lead him to the conclusion that adolescents with more siblings would be more likely to volunteer comments than those with fewer siblings. However, other literature seemed to suggest just the opposite. It contended that only children, who communicated primarily with adults, would be more likely to have acquired verbal skills and would be less intimidated by adults.

The literature left Leon undecided. There appeared to be a certain theme running through the various sources that suggested that the predictor and criterion variables might logically be related. But in which direction? Leon had the additional insight gained from his own (admittedly unscientific) observations. This served to "tip the balance." Finally, he designed a small-scale research study that would test a one-tailed hypothesis:

> Among female adolescents in a treatment group, there is a statistically significant positive correlation between the number of their unsolicited comments and the number of siblings in their families of origin.

Methodology

It was already a common agency procedure to videotape group treatment sessions for use in staff supervision. Thus, Leon had no problem with access to data that he could use to test his one-tailed hypothesis. He received permission readily from the agency administrator and the adolescents to use the videotapes from all his group treatment sessions for his small research study.

Leon operationally defined a case as being a female adolescent who attended at least 75 percent of her group sessions during a four-month period. Having identified 37 clients who met this criterion, he reviewed all videotapes of the sessions with a colleague who was interested in his study. Leon and the colleague developed an operational definition for the variable *unsolicited comments.* An unsolicited comment was judged to have been made only if both Leon and his colleague felt that it met that definition. Next, they recorded the number of these comments that each adolescent made during each group session.

Leon and his colleague totaled the number of unsolicited comments for each case and then divided by the number of sessions she attended. This number provided an average number of *unsolicited comments per session* (criterion variable) for each case. The agency's records provided a means to easily record the predictor variable *number of siblings* for each case. Leon compiled measurements for the two variables into a table similar to Table 8.1.

Findings

Leon used Pearson's *r* to determine if he had any reason to claim support for his one-tailed hypothesis. Using the formula for *r*, he learned that the correlation coefficient between the number of unsolicited comments per session and the number of siblings was .34. Thinking back to the general guidelines that relate to the strength of a correlation, he was somewhat disappointed. However, he also remembered that with larger sample sizes (37 is relatively large for *r*), he did not need a very high coefficient to achieve statistical significance. When he looked at a table for *r* that controls for sample size (Appendix C), he saw that the likelihood of making a Type I error in rejecting the null hypothesis with a correlation of .34 and a sample of 37 was less than .025. (Note that .34 falls to the right of .3246 but is smaller than .3810.)

Conclusions

Leon knew that the .05 significance level is generally accepted as support for a statistical relationship between two variables. He also knew that this meant that if he were to claim a relationship between the predictor and criterion variables, he would be on statistically safe ground. However, as with chi-square, a second step to interpretation is required before support for his hypothesis could be claimed. Was the association in the *direction* that he had hypothesized it would be—that is, a positive association?

Leon recalled that a positive association between two variables (as, e.g., Figure 8.3) means that high values of one variable tend to be found among cases that have high values of the other variable, and vice versa. This meant that, in his data set, members who had high values for the variable *number of unsolicited comments* also should have had high values for the variable *number of siblings,* and vice versa. The data looked the way Leon had hypothesized that they would. He concluded that he had support for his one-tailed hypothesis in the direction predicted.

Leon was realistic about his findings. He knew that there could be other explanations for his statistically significant finding other than that a true relationship between the two variables exists. His research design had been of the ministudy variety. He had, for example, relied on a convenience sample and used only his own cases. Many potential biasing effects and other variables that might have affected his findings could have been present. For example, some bias may have been created within the sample as a result of case loss; Leon may have been unable to be a good facilitator with certain members who were not used to being in group situations; and actual events may have possibly been distorted by the limits of the videotaping equipment used. In addition, the .34 correlation coefficient between the two variables was really not that strong in an absolute sense. His lack of confidence in his findings told Leon that he was not ready to publish an article from his study to communicate his findings to others.

Implications

Leon's findings, even as qualified as they were, were certainly not without value. He summarized them in a weekly staff meeting for other social workers to consider. His colleagues provided a critique of his research methods and identified possible biases and the presence of other rival hypotheses that, if methodologically controlled, would make the design of subsequent research studies even better. All agreed that some replication of his study was certainly indicated.

Leon and a few other social workers decided to make some adjustments to their practice methods based on the assumption that the association that he had tentatively identified was a real one. They perceived little risk to clients in implementing some changes based on the results of his findings. They agreed to evaluate the changes six months later. The social workers decided to take these steps:

1. Staff would use the variable *number of siblings* (the data were available from the intake form) to create more homogeneous groups among new clients. They felt that by placing what might be the most verbally assertive members (those with more siblings) in groups together, they could prohibit them from intimidating other group members who were less assertive. They also hoped that the more assertive clients would be less likely to dominate and monopolize discussion among persons most like themselves. In turn, some of those who the social workers believed to be less assertive (those with fewer siblings) might become more active and assertive in groups with persons more like themselves.

2. In other groups, new members from families with many siblings would be viewed as at risk to try to dominate discussions. Likewise, new members with no or few siblings would be viewed as at risk to be reticent to volunteer comments. This perception would affect the way in which the social worker would approach the role as facilitator with the more heterogeneous groups.

3. In all groups, leaders would facilitate discussion around such areas as attitudes toward the presence or absence of siblings, parental attitudes toward children's assertiveness, and so on.

As in the examples using chi-square discussed in Chapter 7, the contribution to social work's overall body of knowledge offered by Leon's research study must be regarded as limited. The actual amount of knowledge acquired was small and tentative. However, on another level, Leon accomplished a great deal that might benefit himself, his colleagues, and clients. He laid the groundwork for further research studies that would use more sophisticated research designs.

In addition, the social work staff started to think about doing additional research studies. They had begun to apply research findings to social work practice and might in the future attempt to use research findings published in professional journals. Perhaps, without even knowing it, Leon had also moved the staff closer to legitimizing research as an important component of their social work practice. More effective treatment groups might or might not be formed as a direct result of Leon's research study. However, somewhere in the future, his small-scale study might have had some positive effect on the delivery of services to clients in his agency.

MACRO EXAMPLE

Background

Tanya is an administrator in a county department of social services. When she was hired, the error rate for eligibility determinations for new AFDC applications within her agency was among the highest in the state. She assumed that the problem must be related to the inadequate training of the workers. She quickly took steps to increase training for all eligibility workers who had been employed by the agency for less than six months. She also required all senior workers who were not full-time supervisors to perform at least three eligibility determinations each week. To her surprise, one year after she implemented these decisions, the error rate had nearly doubled.

Tanya (and her boss!) was very concerned about the new error rate figures. Tanya wondered how her efforts to address a problem could possibly have made it worse. How could increased training of newer workers, combined with greater use of more experienced personnel, have resulted in a dramatic increase in erroneous eligibility determinations?

In discussing this paradox with a staff member, she began to speculate on what may have gone wrong. The staff member made the casual observation that there had been a series of major changes in federal AFDC eligibility requirements over the past few years. Tanya had asked her senior workers to do more eligibility determinations because she thought they would make fewer errors. However, they were not given additional training to update them on the newer eligibility requirements. Perhaps their greater practice experience, acquired under older policies, was not the asset that she had hoped it would be. In fact, the use of senior workers may have been the major cause of the increased error rate.

Tanya did not wish to make another impulsive decision that might not help the problem—or might make it even worse. If she were to recommend any future changes, she could not rely solely on a logical hunch. She intended to have data to back up future decisions.

Hypothesis

Tanya thought that if experience was really positively correlated with error rate, then she should be able to demonstrate this relationship using data already available in the agency's management information system. Because she was more interested in explaining differences in the variable *error rate* among workers than differences in the variable *amount of experience* among workers, the former was identified as the criterion variable and the latter as the predictor variable. She felt that her brief literature review and conversations with colleagues allowed her to formulate a one-tailed hypothesis:

> Among AFDC eligibility determination workers, there will be a statistically significant positive correlation between years of employment within the agency and error rates.

Methodology

Like Leon in the previous example, Tanya kept her research study simple. Since she needed a quick answer, she limited her study to an examination of the relationship between only the predictor and criterion variables. She calculated a Pearson's *r* to assess the correlation between the two variables among a sample of 42 workers (out of 210) currently doing eligibility determinations. She used the number of identified errors per 100 cases (the last 100 reviewed) for her measurement of the criterion variable.

Findings

The correlation coefficient between the variables *years of experience* and *error rate* was –.21. Tanya checked a table of critical values such as Appendix C to see if chance might have been the explanation for this correlation coefficient. The table told her that she would need a minimum *r* value of .2573 to achieve statistical significance at the .05 level with a sample of 42 and a one-tailed hypothesis. She had not demonstrated support for her hypothesis at the .05 level. What is more, the fact that the correlation coefficient carried a minus sign indicated that the trend was in the opposite direction to that which she predicted she would find. (Technically, she did not have to check the table of critical values to see if her hypothesis had received statistical support since the correlation was in the opposite direction to that which she had predicted.)

Conclusions

Tanya had hoped to find support for her belief that workers with more experience made more errors and those with less experience made fewer errors. She planned to use these findings to recommend to supervisors that the most senior workers should no longer be used for eligibility determinations or that they attend the same training sessions as newer workers. Her findings were a disappointment to her. The slight negative correlation suggested that the most senior workers actually made fewer mistakes than those with less experience.

Implications

Tanya knew enough about the research process to understand that the lack of support for her hypothesis did not mean that no new knowledge had been generated. Her findings helped her to shift her focus away from the past work experience and/or lack of recent training of senior workers as a factor in the recent rise in error rates. She recalled that, in her haste to reduce error rate, she had implemented two changes: Senior personnel were required to perform more eligibility determinations *and* newer staff were given expanded training for their role. Perhaps the problem had been made worse by the introduction of the training for newer staff.

Tanya knew that more training did not guarantee that workers would be better prepared to do their jobs. She began to question whether the training was accomplishing its goals. She decided that she would:

1. Design and implement an evaluative study of the current training for new eligibility workers.
2. Continue to give senior workers increased responsibility for eligibility determinations and encourage them to assist newer workers in learning their jobs.
3. Report her research findings to her superiors, informing them of her approaches to the problem (Steps 1 and 2) and make them aware of her concern about the high error rate and her attempts to correct it.

Tanya's contribution to knowledge using correlation analyses was useful despite the fact that her hypothesis was not statistically supported (and the results were in the opposite direction). Because she used Pearson's *r* correctly, her findings had credibility. They allowed her to approach decision making more knowledgeably and provided guidance for the selection and design of other related research studies.

CONCLUDING THOUGHTS

This chapter has presented correlation as a means of determining and expressing the strength and direction of an association (the extent of covariance) that can exist between variables. We have noted the effect that sample size has on statistical significance and the way that it mathematically explains why a correlation coefficient may be statistically significant while it is actually meaningless.

Correlation is one of many areas in which we must take special care not to deliberately or unintentionally misrepresent our research findings.

Like all statistical tests, Pearson's r does not control for possible rival hypotheses, design bias, or sampling bias. These factors should be controlled through careful attention to research design before data analyses begin. In Chapter 9, we will use the concepts contained within this chapter and shift our attention to simple linear regression. As we shall see, it is highly related to correlation.

STUDY QUESTIONS

1. What can a scattergram portray that other graphs that we have examined cannot? What does each dot in a scattergram depict?

2. Construct hypothetical data on the variables *number of siblings* and *mother's highest school grade completed* for members of your class. Create a scattergram similar to Figure 8.3 to portray your data. Does the scattergram suggest that the two variables are related? If so, in which direction?

3. Why do we use the terms *predictor variable* and *criterion variable* rather than the terms *independent variable* and *dependent variable* in discussing correlation analyses?

4. In a correlation coefficient, what do the sign and the number reflect?

5. What do we mean by linear correlation? Provide examples of variables that might be correlated in a nonlinear way.

6. What does r^2 tell us about the relationship between variables? Provide examples in your discussion.

7. Interpret each of the following hypothetical values for r: $r = -45$; $r = .45$; $r = 1$; $r = -1$.

8. Fill in the blanks to complete each of the following sentences: Correlation coefficients range from _____ to _____. A correlation coefficient suggests the _____ and the _____ of the relationship between two variables. A correlation coefficient of _____ indicates a perfect positive relationship; _____ indicates a perfect negative relationship; and _____ indicates that there is no linear correlation between two variables.

chapter 9

Simple Regression

$\mathbf{A}$s presented in Chapter 8, a correlation coefficient (expressed as an *r* value) gives us an "overall picture" of the relationship between two variables with reference to strength and direction. Another, related form of statistical analysis, simple linear regression, tells us even more about the relationship between two variables. By using simple linear regression, when we know a value, or score, on the predictor variable, we can *predict* (with varying degrees of accuracy) what a value, or score, will be on the criterion variable.

WHAT IS PREDICTION?

In statistical analyses, prediction refers to knowing (without measuring) what a case value, or score, is likely to be for a criterion variable. Understanding statistical prediction involves many of the statistical tools that we have already discussed in the previous eight chapters. For example, the mean and standard deviation (Chapter 3), the normal distribution (Chapter 4), and the correlation coefficient (Chapter 8), all are involved in prediction.

Suppose we want to predict the scores on a standardized life skills measuring instrument following a 12-week life skills training program for people with developmental disabilities. The instrument is administered to each of 25 clients in the program at 6 weeks (midprogram) and 12 weeks (termination). It produces scores that range from 0 (very poor life skills) to 100 (excellent life skills).

To make our predictions, we would want to base our estimates on some fact that we already know. If the only datum that we have is the group's mean life skills score at termination, this mean (e.g., 82) would be our best estimate for predicting a final life skills score for any one client. For instance, if Joe were one of the 25 clients and we wanted to predict his final life skills score, then our best guess would be that he would score 82 at the end of 12 weeks.

Clearly, such predictions based on only a group's mean are rather crude estimates. To improve our prediction, we might look to other data for help. Suppose we knew Joe's score from his 6-week assessment. We could use this datum to make a more informed prediction about what his final score is likely to be at 12 weeks. For example, let us say that Joe's life skills score at six weeks was 79 and that the group's mean score at six weeks was 75. The distribution in Figure 9.1a shows that Joe scored higher than the group's average at the 6-week assessment period. Thus, it would be reasonable to suggest that Joe would also score higher than the group's mean (i.e., 82) at the 12-week assessment period (presented in the distribution in Figure 9.1b).

Without any additional information about Joe or the program, we can statistically improve the accuracy of our prediction by considering the correlation between the scores at the 6- and 12-week assessment periods. As we know from Chapter 8, the correlation coefficient (r) between the two variables helps by telling us (roughly) how accurate our prediction will be. If the correlation is equal to zero, data from the 6-week assessment (distribution in Figure 9.1a) will not be of any help to us in predicting scores at the 12-week assessment (distribution in Figure 9.1b). In other words, we cannot compare data across the two distributions. However, if our correlation is 1.0 or –1.0, then we can make perfect predictions.

Understanding how perfect predictions are made requires knowledge about standard deviations and the normal distribution. Suppose that there is a perfect positive linear correlation ($r = 1.0$) between scores at the 6-week assessment and scores at the 12-week assessment period. Thus, one's position within the group at the 6-week assessment period would correspond perfectly to one's position within the group at the 12-week assessment period. If a client's score was one standard deviation above the group's mean score at the 6-week assessment period, then his or her score at the 12-week assessment period would also be 1 standard deviation above the group's mean score.

Let us take a closer look at Figure 9.1. It shows both the raw scores and the standard scores for the 6- and 12-week assessment periods. The standard deviation for the 6-week assessment period is equal to 4 since the raw scores change by increments of 4 along the base of the 6-week distribution. This being the case, Joe would have scored one standard deviation above the mean for the 6-week assessment on the standard normal distribution (75 + 4 = 79). If the correlation between 6- and 12-week assessment scores was 1.0, we would know that Joe would have the same relative position on both normal distributions. Thus, to predict Joe's score at the 12-week assessment period, we would find the raw score that corresponds to 1 standard deviation. As can be seen from the distribution in Figure 9.1b, Joe's predicted 12-week assessment score would be 87. It is 1 standard deviation above the mean, where $SD = 5$ (82 + 5 = 87).

No matter what the strength of the correlation (we are assuming it is positive), the predicted score for Joe should fall between 82 (our estimate without

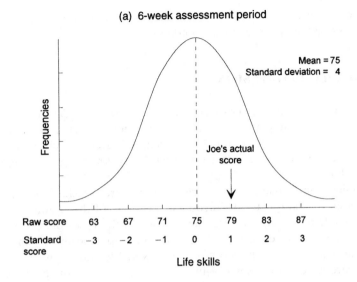

(a) 6-week assessment period

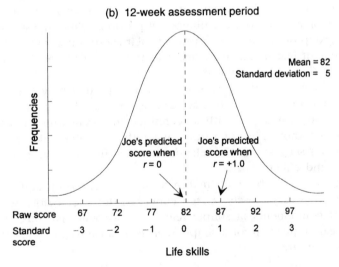

(b) 12-week assessment period

FIGURE 9.1 Standard normal distributions for life skills at 6-week and 12-week assessment periods

any information from the 6-week assessment) and 87 (our estimate when $r = 1.0$ between scores at the 6- and 12-week assessments).

As we suggested earlier, perfect predictions are rare in the real world. Comparing the normal distributions for two variables that have a less than perfect correlation coefficient is not as straightforward as when $r = +1.0$ or -1.0. When the correlation between two variables is less than perfect, we must rely on simple regression analysis to help us to make predictions.

WHAT IS SIMPLE REGRESSION?

Simple regression is also known as *simple linear regression*. It can be used when two variables are correlated to some degree and they reflect a linear relationship when plotted on a scattergram. Simple regression allows us to make predictions about the value of a criterion variable by knowing the value of a predictor variable.

Simple regression is useful in experimental research designs in which the researcher is able to manipulate or introduce the value of the predictor (technically, the independent) variable. It also can be used in the types of research situations we more commonly encounter in social work. Specifically, it can be used to examine the relationship between a predictor variable and a criterion variable when the researcher can only select—but cannot manipulate or introduce—the predictor variable. For example, we could use simple regression to examine the predictive value of a student's undergraduate grade point average (GPA), the predictor variable, on the student's GPA in graduate school, the criterion variable. Or we could use it to look at how well the number of previous felony arrests (predictor variable) for a group of offenders can predict their scores on a standardized self-report measurement instrument that measures the degree of hostility they have toward authority (criterion variable).

In either example, simple regression is used to predict how well a person (or case) will "perform" on a particular variable that can be measured. It does nothing more than provide us with a technique for predicting values of one variable given our knowledge of values of another variable. Like correlation, regression does *not* suggest that the relationship identified between two variables is one of cause and effect.

Simple regression can be explained in terms of a regression equation and a regression line—two very closely related concepts. However, before we discuss these important components of simple regression, we will first look at how to formulate a research question for a simple regression analysis and what the limits of simple regression are.

Formulating a Research Question

Simple regression relies heavily on the strength and statistical significance of a correlation coefficient (r) to make predictions about values of the criterion variable. It is not necessary to restate a null and research hypothesis when using simple regression. The null hypothesis—that there is no relationship between the criterion and predictor variable—would already have been tested when Pearson's r was computed.

Instead of restating a research hypothesis, researchers who use simple regression typically specify a research question. A research question for simple regression analysis asks how well knowing a value of the predictor variable can improve prediction on the criterion variable. The following are possible research questions (based upon our previous examples) that can be answered using simple regression analysis:

How much does knowing a student's GPA in undergraduate school help in predicting his or her GPA in graduate school?

How well does knowing the number of previous felony arrests contribute to predicting a felon's degree of hostility toward authority as measured by a standardized self-report instrument?

Limits of Simple Regression

A simple regression analysis produces an equation, called a *regression equation.* Using the equation, we can predict (with less than total accuracy) the value of a criterion variable (*Y*) for a particular case by knowing the value for the case's predictor variable (*X*). This would be possible even for those cases whose measurements of the predictor and criterion variables were not actually used in the formula to develop the equation. However, there are limits to what we can legitimately do using simple regression. We cannot make predictions using values for the predictor variable that are either larger than the largest *X* value or smaller than the smallest value used in the computation of the equation. For example, suppose that the predictor variable—number of hours of vigorous exercise per week—was used to predict longevity. If our sample (on which the regression equation was computed) included individuals who exercised between 2 and 15 hours per week, then we could not use the equation to predict longevity for other individuals with less than 2 or more than 15 hours of vigorous exercise per week.

Other statistical limitations of simple linear regression are the same as for Pearson's *r.* Both types of analysis require the same level of data (interval or ratio) and rest on the same statistical assumptions (see Chapter 8).

COMPUTATION OF THE REGRESSION EQUATION

Let us use an example to show how to compute a simple regression equation. Suppose that we wish to find out whether, among a group of adolescent female clients, there is a relationship between their educational level (predictor variable) and their assertiveness (criterion variable). Their educational level is measured by recording the last year of school completed, and their assertiveness is measured by the completion of a self-report standardized measuring instrument that measures assertiveness.

Simple regression can help us predict one adolescent's assertiveness score (criterion variable) by just knowing her educational level (predictor variable), provided the two variables are found to reflect a linear correlation. Suppose the seven

TABLE 9.1 Educational and assertiveness
levels (*N* = 7)

Name	Educational Level (X)	Assertiveness Level (Y)
Rochelle	9	23
Carny	12	29
Belinda	6	17
Amanda	8	21
Maria	11	27
Ky	15	35
Ruth	13	31

clients selected for study reflect the two distributions (i.e., educational level and assertiveness level) presented in Table 9.1.

If we study the data in Table 9.1, we can make a number of observations. For all cases, the measurement of *Y* (the criterion variable, or assertiveness level) is greater than *X* (the predictor variable, or educational level). It is more than twice as great. In fact, in all seven cases, *Y* is exactly 2(*X*) + 5. Expressed as an equation, it would be:

$$Y' = 5 + 2(X)$$

If we were to plot the data in Table 9.1 using a scattergram and connecting the dots as in Figure 9.2, we would have a straight line. For example, Rochelle has 9 years of education (*X*) and has an assertiveness level of 23 (*Y*). Thus, the regression equation reflects the relationship between her education level and her assertiveness score: $Y' = 5 + 2 (9) = 23$. The equation $Y' = 5 + 2(X)$ holds true

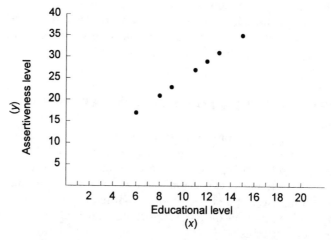

FIGURE 9.2 Scattergram of educational and
assertiveness levels (from Table 9.1)

for the other six adolescents as well. This is because the data form a straight line—a rarity in the real world.

The regression equation for the data in Table 9.1 and Figure 9.2 ($Y' = 5 + 2[X]$) is a variation of the equation for a straight line, which is $Y = a + b(X)$. The general regression equation when we wish to predict a value of the criterion variable is:

$$Y' = a + b(X)$$

Where:

    Y' = the predicted Y value from a particular X value.
    a = the point where the regression line intersects the y-axis, the constant.
    b = the slope of the line where the amount of change in Y is directly related to the amount of change in X, the regression coefficient.
    X = a selected value of the predictor variable used to predict the value of the criterion variable.

Hand computation of a regression equation would be a lengthy process, especially for a large data set of, say, 1,000 cases. However, computers can do the computation effortlessly in very little time. Most statistical software packages are programmed to display the data within a scattergram showing dots for all cases, the correlation coefficient (r), and the regression equation. For those who may wish to see how a regression equation is calculated using formulas, Box 9.1 has been provided. It uses the data displayed in Table 8.2 in Chapter 8.

THE REGRESSION LINE

The regression line is a straight line of "best fit" for a given set of data and can be presented mathematically (as above) or graphically in a scattergram. The values in Table 9.1 as illustrated in Figure 9.2 reflect a perfect positive correlation ($r = 1.0$), but, as we repeatedly have emphasized, perfect positive correlations (and negative ones as well) are very rare, especially in social work research. It is far more likely that the actual data in our example would look more like those in Table 9.2 and Figure 9.3, which reflect a less than perfect linear correlation between the two variables of educational level and assertiveness level.

A straight line cannot be drawn through the seven dots in Figure 9.3, since the correlation between the two variables displayed is not a perfect one (it is neither +1.0 or -1.0). A computer printout would tell us that $r = .76$ for the two variables in Figure 9.3. We could draw several lines through what appears to be the center of the distribution of dots, but it would be hard to determine which one would do the best job. Some lines would be more vertical or more

BOX 9.1
The Least-Squares Regression Equation

Problem: What is the least-squares regression equation ($Y' = a + b[X]$) for the data in Table 8.2?

Predictor Variable: Unemployment rate—ratio level.

Criterion Variable: Number of civil disturbances—ratio level.

Data: See Table 8.2.

Slope Formula:
$$b = \frac{N\Sigma XY - (\Sigma X)(\Sigma Y)}{N\Sigma X^2 - (\Sigma X)^2}$$

Where:
b = Slope
N = Number of cases
ΣXY = Sum of xy column
ΣX = Sum of x column
ΣY = Sum of y column
ΣX^2 = Sum of x^2 column
ΣY^2 = Sum of y^2 column

Substituting values for letters:

$$b = \frac{(5)(985) - (76)(53)}{(5)(1290) - (76)^2}$$

$$= \frac{897}{674}$$

$$= 1.33 \text{ (slope)}$$

y-intercept Formula:
$$a = \bar{Y} - b\bar{X}$$

Where:
a = y-intercept
$\bar{Y}$ = Mean of Y column
$b\bar{X}$ = Slope times mean of X column

Substituting values for letters:

$$a = \left(\frac{53}{5}\right) - (1.33)\left(\frac{76}{5}\right)$$

$$= 10.6 - (1.33)(15.20)$$

$$= 10.6 - 20.2$$

$$= -9.62 \text{ (y intercept)}$$

Presentation of Results: $Y' = -9.62 + 1.33(X)$

Conclusion: A slope of 1.33 means that for every unit change in X (for every increase of 1 in the unemployment rate) there was a change of 1.33 units in Y (the number of civil disturbances increased by 1.33). Note from Box 8.1, $r = .82$.

TABLE 9.2 Educational and assertiveness levels ($N = 7$)

Name	Educational Level (X)	Assertiveness Level (Y)
Rochelle	9	20
Carny	12	21
Belinda	6	14
Amanda	8	25
Maria	11	21
Ky	15	28
Ruth	13	30

horizontal than others. We would want find the line with the best possible fit— one, and only one, line—that would pass through "the center of the dots," hitting as many as possible while not missing the others by too much. What would be the slope of such a line?

The Least-Squares Criterion

The line that we would seek for the data in Figure 9.3, its regression line, is a line of predicted scores (Y'). As we know, the dots above and below the line represent actual scores. Notice how the actual scores above the line are predicted

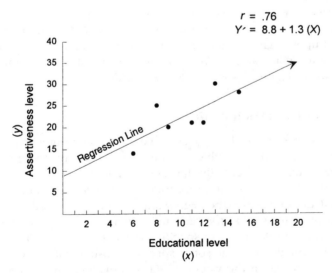

$r = .76$
$Y' = 8.8 + 1.3 (X)$

FIGURE 9.3 Scattergram of educational and assertiveness levels (from Table 9.2)

by the line to be lower than they actually are and the scores below the line are predicted by the line to be higher than they actually are. This is basic to an understanding of regression.

The best way to obtain a regression line, that is, to determine its equation reflecting its slope and where it would intercept the y-axis, is to use what is referred to as the *least-squares criterion*. Obviously, any line drawn through the center of the dots in Figure 9.3 would miss some of the dots. However, the distance between each of these respective dots and the line could be measured. Then each of these distances (referred to as *deviations*) could be squared and their squares added together. The sum of the squared deviations from a regression line would be different for each different line we draw through the dots. The least-squares criterion asserts that the best line, the regression line that *should* be used, is the one that produces the smallest sum of squared deviations from that line.

Fortunately, there is a much easier and better way to arrive at a regression line for a data set than by trying an almost infinite number of lines, determining their sum of squared deviations and then selecting the line with the "least squares." It involves the computation of the regression equation. We have already seen (Box 9.1) that a mathematical formula is available to calculate the slope and the point of the y-intercept for any pair of interval, or ratio level, variables. The formula generates a linear equation for the regression line that we are seeking. Both the a value (where the line intercepts the y-axis) and the b value are likely to be fractions and not whole numbers, as in our first example where a perfect correlation existed.

The regression equation uses the notation Y' (Y prime) to denote the fact that what the regression formula produces is a *predicted* value of the criterion variable based upon a given value of the predictor variable (X), not what it was observed to be using actual data. The computer-generated equation for the regression line shown in Figure 9.3 is $Y' = 8.8 + 1.3(X)$.

The Regression Coefficient (b)

The first regression statistic that is calculated by the computer is referred to as the regression coefficient (b), which is the slope of the regression line. (*Slope* as used in this book has essentially the same meaning as it does when we refer to the slope of a hill—that is, the proportional relationship between vertical rise and horizontal distance.) For the regression line, this is stated as the amount and direction of change in Y' for each unit change in X. If b is positive, as X increases, Y' will increase at some specific constant rate. Or, if b is negative, as X increases, Y' will decrease at some specific constant rate. The b value suggests the rate of change in the values of Y (the criterion variable) for each unit of change in the value of X (the predictor variable).

The slope of the regression line (b) indicates exactly where (between the vertical and horizontal axes) the line would fall. High b values reflect lines that are nearly vertical; low b values reflect lines that are nearly horizontal.

The regression coefficient (*b*) is closely related to the correlation coefficient (*r*) for the two variables. When $r = 0$, $b = 0$, and when $r = +1.0$ or -1.0, *b* reaches its maximum value. Unlike the value of *r*, which always ranges between $+1.0$ and -1.0, the value of *b* may be of any size. Also, as we have suggested, *r* and *b* will always correspond in terms of positive and negative values. If *r* is a negative value, *b* also will be negative; if *r* is positive, *b* also will be positive.

Computer printouts that give us *b* also give us *r*. We can determine the probability that any apparent relationship between the predictor and criterion variables (as expressed by the regression coefficient) may have occurred by chance. For a two-variable equation, the significance level for *r* will be exactly the same as the significance level for *b*.

After the data are entered into a computer and a regression analysis is completed, the computer printout will display the value of *b* for the two variables. The *b* value for Figure 9.3 is 1.3; it is seen in the upper righthand corner of the figure. Since this is a positive number, this means that for each unit increase in *X*, *Y'* increases 1.3 points. Thus, for each one-point increase in educational level, we can predict an increase of 1.3 points in "assertiveness level."

Y-Intercept (*a*)

After the computer has determined the slope of the regression line, it will also display the starting point for the line. The starting point is the *y*-intercept, the point at which the regression line crosses the *y*-axis. This is the point that represents the predicted value of *Y* when *X* equals zero. In our example, the equation would be $Y' = 8.8 + 1.3(0) = 8.8$. It is also the point that represents *Y'* when $b = 0$. In this case, the equation would read, $Y' = 8.8 + 0(X) = 8.8$.

Computer statistical software draws the correct regression line in a scattergram and also shows the place where it crosses the *y*-axis. The value of *a* can be positive or negative. A positive *a* value tells us the regression line will meet the *y*-axis at a value *above* zero; a negative *a* value indicates the regression line will cross the *y*-axis at a value *below* zero.

Predicted Y (*Y'*)

The regression coefficient (*b*) and the *y*-intercept (*a*) help us to generate a regression line. *Y'* (the regression line) is the estimated score for the criterion variable, based on a given value of the predictor variable. When plotted on a scattergram, the predicted values of *Y* will always form a straight line.

With the regression equation formulated and the regression line displayed, it is possible to make predictions for a single case by computing *Y'* for different values of *X*. Look at the upper righthand corner of Figure 9.3 for the regression equation $Y' = 8.8 + 1.3(X)$. The *X* component of the formula is *any* specific value of *X* for which we wish to predict a *Y* value. The following are calculations for the predicted *Y* values of two different cases when we substitute specific values for *X* (13 and 6, respectively):

Calculation 1:
Let us determine Y' by substituting 13 for X:
$Y' = a + b(X)$
    $= 8.8 + 1.3(13)$
    $= 8.8 + 16.9$
    $= 25.7$ (predicted value of Y when $X = 13$)

Calculation 2:
Let us determine Y' by substituting 6 for X:
$Y' = a + b(X)$
    $= 8.8 + 1.3(6)$
    $= 8.8 + 7.8$
    $= 16.6$ (predicted value of Y when $X = 6$)

The X values of 13 and 6 were chosen at random. They could have been any numbers within the range of the distribution X. What do the X and Y' values mean? When X is 13, the predicted Y value is 25.7 (Calculation 1); when X is 6, the predicted Y value is 16.6 (Calculation 2). Thus, a client who has an educational level of 13 is predicted to have an assertiveness score of 25.7, and a client with an educational level of 6 is predicted to have an assertiveness score of 16.6.

An alternative approach to predicting the value of Y involves using the actual regression line as presented in a scattergram. For example, we could find the exact value ($X = 13$) along the x-axis of the scattergram in Figure 9.3. Next we could run a vertical line straight up until we reach the regression line. Then we could draw a straight horizontal line to the left until we touch the y-axis. Note that we hit the same Y' as predicted in the equation above (Y' = about 25.7). This method of prediction is not nearly as precise as using the regression equation. However, it does provide a visual understanding of the data and can be used as a quick check on a result obtained mathematically. We could use the same method to determine the value for Y' when $X = 6$.

INTERPRETING REGRESSION ANALYSIS

We have conceptualized years of education as the predictor variable (X) and degree of assertiveness as the criterion variable (Y). We could have conceptualized the relationship between the variables the other way around, however. Assertiveness could be thought of as the predictor variable and years of education as the criterion variable. As we emphasized earlier, the labeling of variables as predictor and criterion variables needs to make theoretical and logical sense. In this instance, we could build a logical case for a relationship between variables in either direction.

The predictor and criterion variables are not interchangeable for statistical analysis purposes, however. If we were to tell the computer that Variable X is the predictor variable and Variable Y is the criterion variable, we would get one result (regression equation). If we were to tell the computer that Variable Y is the predictor variable and Variable X is the criterion variable, we would get a different result. We must be clear as to what variable is intended as the predictor variable and what variable is intended as the criterion variable (as stated in the research question), and we must be careful to input these data correctly into the computer when a simple regression analysis is to be performed.

Presentation of Y'

Once we have decided which variable will be the predictor variable and which the criterion variable, the presentation of Y' is straightforward. For example, the regression equation $Y' = 3.5 + 6.8(X)$ could be stated in words as "for every unit increase in X there will be a 6.8 unit increase in Y'. Using the variables of client's educational level (X) and assertiveness level (Y), this would read, "for each one-year increase in a client's educational level, a client's assertiveness level would increase by 6.8 assertiveness points."

Meaningfulness of Y'

An obvious question always comes to mind when trying to predict with simple regression. That is, how much advantage do we gain by making predictions from the regression line? The answer depends on the strength and significance of the correlation between the two variables. When there is very little correlation, there is no use in calculating the best-fitting straight line to make predictions. However, when there is a high correlation, it pays to use this line for predictive purposes. Thus, there is a direct relationship between the magnitude of r and the degree to which our predictions can improve accuracy in prediction.

If we study the formula in Box 9.1, it clearly shows how correlation is directly related to regression. As r decreases (goes to 0.0), the accuracy of prediction using regression decreases. As r increases (goes to +1.0 or -1.0), the accuracy increases.

Standard Error

The standard error of an equation gives us an estimate of how well the regression equation can predict values of the criterion variable, from values of the predictor variable. A 100 percent accurate prediction is possible using simple regression only when the correlation between the predictor and criterion variables is perfect. When $r = +1.0$ or -1.0, all dots will fall right on the regression line. The standard error, also known as the standard error of estimate, will equal zero.

The smaller the standard error of an equation, the less scattered the scores will be around the regression line. In other words, the closer the standard error is to zero, the more confident we can be that the regression equation will give us good predictions. The further r is from being perfect, the more we must account for error in the equation. When $r = 0$, the standard error will be at its maximum.

MICRO EXAMPLE

Background

Wilhelmenia is a social worker in a family service agency. Over the last year she saw 80 clients who were undergoing treatment for depression. During this time she formulated the impression that the clients who seemed most depressed at intake appeared overwhelmed by the little amount of free time they had to themselves because of all their day-to-day responsibilities. She also felt that clients who had more free time during the week seemed less depressed than those clients who had little free time. Wilhelmenia decided to do a small-scale research study to test her hunch.

Research Question

Wilhelmenia wanted to determine how well information about the amount of client free time would improve her ability to predict a client's depression level. Her predictor variable was the average amount of free time a client has during the week (X), and the criterion variable was client depression level (Y).

Methodology

Over the next several months, Wilhelmenia had each of 50 new clients who were about to be seen for depression complete a standardized self-report measurement instrument that measures depression (interval level). She also asked each one how many hours of free time, on the average, he or she had during the week (ratio level). The depression instrument produces scores ranging from zero (highest level of depression) to 200 (lowest level of depression).

Findings

Wilhelmenia entered the data for the two variables into her personal computer. She knew that the regression equation that would be produced by her data would be derived from the formula for a straight line:

$$Y = a + b(X)$$

The Pearson's r and regression equation that the computer produced for her data were:

$$r = .70$$
$$Y' = 1.49 + 4.69(X)$$

Conclusions

Wilhelmenia studied the above findings. She concluded that a slope (b) of 4.69 indicates that, for each unit change in X, there is an increase of 4.69 units in Y'. For example, for each additional hour of free time a client has during the week, there results an increase of 4.69 "points" on the depression instrument (a decrease in depression). To be sure her prediction equation was credible, Wilhelmenia made certain that r was statistically significant, which it was. She also noticed that the standard error was small, which suggested that the regression equation could generate worthwhile predictions. Finally, she computed r^2 and decided that the amount of clients' free time was related to a fairly sizable amount of the variation of their depression levels ($.70^2 = .49$, or 49 percent).

Implications

Wilhelmenia could use her findings to predict a future client's depression level just by knowing how many hours of free time the client had during the week. She realized that her findings are not very generalizable to other clients with different problems (i.e., those who do not have a presenting complaint of depression) or to other clients being treated for depression in other settings. She also knew that she could not predict client success with future clients who had more or less free time during the week than any contained in her original 50-case data base.

Wilhelmenia also realized that there was great potential for design bias in her study. In addition, it would be nearly impossible to eliminate all rival hypotheses that could account for the relationship between the two variables (e.g., number of children, income level, single or double parenthood). However, she could now predict, with some degree of accuracy, a value of Y' by knowing a value of X—the purpose of simple regression.

After completing her study, Wilhelmenia saw a 51st client, David, who came to the agency to be treated for depression. During their first interview, David told Wilhelmenia that he had an average of 15 hours of free time during the week. Wilhelmenia computed David's predicted depression score (Y') as follows:

$$Y' = 1.49 + 4.69(15)$$
$$= 1.49 + (70.35)$$
$$= 71.84 \text{ (predicted value of } Y \text{ when } X \text{ equals 15)}$$

To test her prediction equation, Wilhelmenia had David complete the depression measurement instrument and obtained his actual score. She compared his actual score ($Y = 73$) to his predicted score ($Y' = 71.84$) and was pleased to find that they were quite close.

MACRO EXAMPLE

Background

Miriam is an administrator in a child protection agency. With recent cutbacks in funding, caseloads have increased for Miriam's workers. In the past, her child protection workers typically managed a caseload between 30 and 40 clients. More recently, workers were assigned to 40 to 50 cases at a time. Miriam noticed that since caseload size had increased, there also seemed to be an increase in sick days taken by her workers. She wondered if the size of a worker's caseload might be a good predictor of the number of sick days a worker might take.

Research Question

Implicit in Miriam's thinking was the hypothesis that caseload size and number of sick days are positively correlated (a one-tailed hypothesis). In other words, as caseload size increases, number of sick days also increases. The predictor variable for her simple regression analysis was *caseload size,* and the criterion variable was *number of sick days.* Both variables are at the ratio level of measurement.

Methodology

To test her theory, Miriam reviewed employee records over the past year. There were 26 front-line child protection workers on staff. She eliminated from her analysis two workers who have been receiving ongoing medical care for chronic health problems. She did this because she feared that they would somehow bias the results of her ministudy. She was left with data from 24 cases to examine. She reviewed the 24 employee records and randomly selected a one-month interval for each worker. She noted the workers' caseload size and the number of sick days taken during that month.

Findings

After entering the data into her personal computer, Miriam performed a simple regression analysis. She remembered that the regression equation assumes that caseload size and number of sick days have a linear relationship. In other words, when the two variables are plotted on a scattergram, a relatively straight line should be detected.

Miriam examined her computer printout and noted the following:

$$r = .76$$
$$Y' = -6.8 + .25(X)$$

Conclusions

Miriam assessed the standard error to determine how well the regression equation can predict the number of sick days, given information about caseload size. She also checked the p value for r to be sure that the relationship between the two variables was unlikely to be the work of chance. Based on her findings, Wilhelmenia was confident that her regression equation could generate reasonably accurate predictions and that the association between caseload size and number of sick days was real and quite strong. The direction of the relationship between the two variables was also positive, exactly what she had expected. She noted that the regression coefficient (b) of .25 meant that for every unit change in X, there is an increase of .25 in Y. In other words, for every additional case added to a worker's caseload, it increases the number of sick days a worker is likely to take by .25.

Implications

Miriam could use her findings to predict the number of sick days that a worker might be expected to take in a month. However, she knew that her findings might be useful for prediction only among those workers in her agency.

After completing her regression analysis, Miriam examined the current caseloads of her staff. She noticed that for most workers, caseload sizes were higher than normal, so she expected that she would have a high number of sick days that month. Wilhelmenia predicted how many sick days a worker with a caseload of 43 would be likely to take for the month by using the following equation:

$$Y' = -6.80 + .25(43)$$
$$= -6.80 + (10.75)$$
$$= 3.95 \text{ (predicted value of } Y \text{ when } X \text{ equals } 43)$$

Wilhelmenia anticipated that a worker with a caseload of 43 would take almost four days of sick time during the month. In efforts to preserve quality service to clients, Miriam immediately reviewed her budget to see if she could hire a recently retired worker on a part-time basis to cover cases of those workers on days that they would call in sick. She did not want to overburden her "healthy" workers with additional work. On a more proactive note, Miriam also decided to call in those workers with the highest caseloads to discuss stress management strategies for the month ahead.

CONCLUDING THOUGHTS

This chapter presented an overview of simple regression analysis and demonstrated how it is closely related to correlation. Regression provides us with a technique for predicting the exact values of one (criterion) variable given our knowledge of values of a second (predictor) variable. It is a useful technique for those situations where a reasonably strong linear correlation can be shown to exist. However, when there is little or no correlation between two variables, creation of a regression equation can be an exercise in futility—it does little to enhance our ability to predict values of the criterion variable by knowing the value(s) of a predictor variable.

STUDY QUESTIONS

1. What can simple regression do that correlation cannot do? Provide examples in your discussion.

2. What statistical insights discussed in earlier chapters in this book are critical to an understanding of simple regression? Describe how they relate to it.

3. Explain why it is not necessary to state a research or null hypothesis when using simple regression. What alternative to restating an hypothesis do researchers commonly use when performing simple regression analysis?

4. Discuss (using examples drawn from social work practice) why it is not a productive use of one's time to use regression when the correlation coefficient between two variables is low. Draw two scattergrams to illustrate your point (one scattergram showing a low correlation and one scattergram showing a high correlation). Discuss the two scattergrams in relation to simple regression.

5. What do we mean when we say that the predictor and criterion variables are not interchangeable for statistical purposes when we perform simple regression? What implications does this have when we instruct a computer to conduct a simple regression analysis?

6. In your own words, explain why you think we should not predict Y' when an X value is below the minimum or above the maximum value used in generating Y'.

7. Describe two situations where you believe it would be appropriate to perform regression analysis in a human service agency with which you are familiar. How would the results have the potential to help the agency?

8. For each of the two major examples used in this chapter (micro and macro), discuss additional methodological limitations (in addition to those mentioned) that might serve to "depreciate" the statistical findings. You may wish to go to a book on social work research methods and review the section on threats to internal and external validity.

chapter **10**

t Tests and *ANOVA*

One of our profession's greatest needs is for research studies that evaluate the effectiveness of our intervention methods. Social workers need to know whether a particular intervention is associated with positive (or negative) changes in client attitudes or behaviors. In addition, we often need to know if one intervention method is better than another to help produce a desired change.

In testing hypotheses related to practice effectiveness, we frequently have reasonably precise measurements of the dependent variable. Examples of dependent variables may be *client self-esteem, public attitudes toward welfare recipients, client marital satisfaction,* or *incidence of spouse abuse.* Thanks to the work of researchers over a period of many years, measuring instruments used to measure these variables are now assumed to generate interval level data.

Other dependent variables, such as *number of appointments missed, number of stated oppositions to a proposed social program,* or *amount of illegal drugs consumed,* are at the interval or ratio levels by their very nature. Many of these variables also tend to be normally distributed within the population, making the mean an appropriate measure of their central tendency and the variance and standard deviation appropriate indicators of their variability (see Chapter 3).

In testing research hypotheses related to practice effectiveness, the independent variable is likely to be considered at the nominal level of measurement. It may be different types of treatment interventions, race or gender of the social worker, or some other measurement that similarly reflects only different categories of the variable. Also, the research sample (or samples) available for study may be relatively small; large samples are not common in social work research, particularly in clinical situations.

The three conditions described above—one nominal level independent variable, one normally distributed interval or ratio level dependent variable, and a relatively small sample—are ideally suited to a statistical analysis that involves a comparison of two or more means. This chapter presents a "family" of related statistical tests used for this purpose called *t* tests and analysis of variance.

t TESTS

Very often in social work research, the dependent variable is measured within either a single sample or within two samples. In a single-sample situation, we may wish to know if the research sample really differs greatly from the population from which it was drawn in relation to the dependent variable—that is, how typical is the research sample? In the two-sample situation, we may wish to know if measurements of the dependent variable differ greatly between the two samples in relation to the dependent variable—that is, are the two samples really all *that* different? Three different forms of *t* tests are used to answer these questions: the one-sample *t* test, the correlated groups *t* test, and the independent groups *t* test.

One-Sample *t* Test

The *one-sample* t *test* compares the mean of an interval or ratio level variable within a research sample with the mean of that variable for the population from which the sample was drawn. The formula for the test produces a *t* value that then is compared with a table like Appendix D to get a *p* value. More commonly, when a computer is used, we get both the *t* value and the corresponding *p* value as products of the data analysis, along with the degrees of freedom. For the one-sample *t* test, the degrees of freedom are the number of cases in the sample minus one.

There are two common uses for the one-sample *t* test. The first relates more to research design and only indirectly to our focus on hypothesis testing, while the second relates more directly to it. The first way the one-sample *t* test is used is to evaluate the representativeness of a research sample in regard to some variable. We may want to know how similar the research sample is to the population from which it was drawn. When used this way, researchers generally hope to *not* reject the null hypothesis. They hope to generate support for the position that any differences between the mean of the research sample and the mean of the population are so small that they probably are the work of chance—that is, that the research sample is indeed representative of the population from which it was drawn in regard to the variable.

We will use an example to demonstrate the first use of the one-sample *t* test. Suppose we want to conduct a research study that attempts to examine the relationship between age and job satisfaction among AFDC caseworkers. We have been told by the program's administrator that we can draw only a small random sample ($N = 40$) of caseworkers since we need to conduct extensive interviews with them. We would like our research sample to be representative of the population of all AFDC caseworkers in relation to the variable *size of caseload.* We know from agency statistical data that the caseload for all AFDC workers is normally distributed and has a mean of 75.

We could draw our research sample of 40 and then, using the one-sample *t* test, compare its mean caseload with the population mean caseload of 75. If the results of our statistical analysis were to suggest that the difference between the two means was relatively small and likely to be the work of chance ($p >$ .05), we would not reject the null hypothesis. The sample would be "not all that different," probably close enough to be considered representative of the population from which it was drawn in regard to the variable *size of caseload.*

If, however, a one-sample *t* test analysis compared the mean of the sample with the mean of the population and concluded that the likelihood of producing two means that were that different, given the sample size of 40, was less than 5 percent ($p < .05$), we would reject the null hypothesis. We might conclude that the sample and the population really *are* different in regard to size of caseload. We would then have to conclude that the research sample is not representative of the population in relation to this variable. If we were to continue with our research study using the sample, it might jeopardize our ability to generalize the study's findings to the population from which the sample was drawn.

The second use of the one-sample *t* test uses the test in a similar way but has the opposite objective. It attempts to gain support for a research hypothesis (one or two tailed) by demonstrating that a sample is so different from its population in regard to some interval or ratio level variable that the difference is not likely to be the work of chance. It thus seeks to be able to *reject* the null hypothesis. We will offer another example to illustrate how this second use of the one sample *t* test might be used.

Suppose, using AFDC caseworkers again, that we hope to demonstrate statistical support for our belief that AFDC workers in our county agency do better overall on the state licensure examination than AFDC caseworkers in other counties. Our one-tailed research hypothesis might state, "Social workers employed at ABC county office will score higher than their peers on the state licensure examination." We could collect the scores from a random sample of 30 of our workers who took the examination during the past five years and, using the one-sample *t* test, compare their mean score with that of all AFDC workers who took the examination during the past five years. Perhaps the mean score of our sample turned out to be higher than the mean score of the population. That alone would not be sufficient to lend support to the research hypothesis. The question remains, was it *that* much higher? Would we be safe in concluding that the higher mean score of our sample was not just the work of chance?

The results of a one-sample *t* test analysis would tell us. It would result in both a *t* value and a corresponding *p* value, which would be based on the *t* value and degrees of freedom. If we had used .05 as our rejection level, a *p* value less than .05 ($p < .05$) would offer support to the decision to reject the null hypothesis and would lend support for the research hypothesis. A *p* value greater than .05 ($p > .05$) would support a decision not to reject the null hypothesis. It would suggest that the difference between the mean score of the sample and the mean score of the population is probably just the work of chance.

Presentation.   The presentation of one-sample *t* tests is straightforward. Table 10.1 provides a clear example of how this is done using the previous example. It gives the results from a hypothetical study that had this one-tailed hypothesis: "Social workers who work in ABC county office will score higher than their peers on the state licensure examination."

The most frequently used presentation of the one-sample *t* test entails displaying the population mean (80), the mean of the research sample (90), the standard deviation for the research sample and for the population (5 and 6, respectively), the number of cases in the research sample and in the population (30 and 100, respectively), the *t* value (2.62), the degrees of freedom (the sample *N* minus 1 = 29), and the corresponding *p* value (< .01).

By glancing at Table 10.1 one could conclude that the 30 workers who took the examination in ABC county had a mean score of 90 compared to the mean score of 80 for the entire population, and that the workers (N = 30) in ABC county thus scored, on the average, 10 points higher than the population (N = 100) as a whole. This 10-point difference is statistically significant at the .01 level (one-tailed test). This means that a 10-point difference would happen by chance less than 1 time out of 100 with this size sample.

Correlated Groups *t* Test

Both of the other two forms of *t* tests also involve a comparison of means. But unlike the one-sample *t* test, they do not compare the mean of a sample with the mean of a population. These two tests use two samples, which may be unequal in size, and they compare the means of these two samples with each other. They ask the question, "Is the mean of one sample different enough from the mean of the other that we would be safe in concluding that there is a real difference between the two samples in relation to the dependent variable?" While both tests compare the means of two samples, they have slightly different formulas.

The *correlated groups* t *test,* also referred to as the paired groups or matched groups test, is used when the research participants in the two samples are connected or matched in some way. For example, the two samples may con-

TABLE 10.1   State licensure examination scores broken down by offices

Subjects	Mean	Standard Deviation	*N*
Sample	90	5	30
Population	80	6	100

t = 2.62, *df* = 29, *p* < .01

sist of pairs of siblings that are assigned one to each sample or of pairs of research participants matched on the basis of psychiatric diagnosis and assigned in the same way. The correlated groups *t* test can be used when the two samples are the same people measured at two different times—in this case, the two "samples" are really just two different measurements of the same people taken at two different times.

The latter use of the correlated groups *t* test is especially useful for social workers who wish to know if some intervention seemed to "make a difference." It can be used with a one-group pretest-posttest research design where only one sample is available for study. For example, we could draw a random sample of clients who smoke and ask them to record the average number of cigarettes they smoked a day before enrolling in a smoking cessation program. We could then record the number of cigarettes they smoked the day after they complete the program. We now compare the mean number of cigarettes the group smoked one day before the program (Sample 1) with the mean number they smoked one day after the program (Sample 2).

The correlated groups *t* test produces a corresponding *p* value that would allow us either to claim a statistically significant relationship between having been in the group and average number of cigarettes smoked ($p < .05$) or not to claim it ($p > .05$). Of course, even if the relationship were found to be statistically significant, we could not reject the null hypothesis that the intervention makes no real difference on this basis alone. We would also have to rule out the possibility that rival hypotheses, design bias, or sampling bias may have created the apparent relationship. For example, the time lapse between the date the data for the first sample were collected and the date the data for the second sample were collected (and what may have occurred for our participants in the interim) may be a more plausible explanation than the intervention for any reduction in cigarette smoking.

Presentation.   The presentation of the correlated groups *t* test is simple. Table 10.2 provides the results of a hypothetical research study with this one-tailed research hypothesis: "Pregnant teenagers will have increased child-rearing skills after they undergo a child-rearing skills training workshop." A five-day child-rearing skills training workshop was offered to 22 pregnant high school students. The group's mean child-rearing skills score was calculated before the skills training workshop (pretest) and after it (posttest).

Table 10.2 displays the results of the study using a correlated groups *t* test analysis. Notice that the students had a pretest mean score of 4.8 and a posttest mean score of 5.8. The difference between the two means was 1 point (5.8 − 4.8 = 1). This 1-point difference is statistically significant at the .01 level with 21 degrees of freedom (*N* minus 1 for this *t* test). This means that a 1-point difference would happen by chance less than 1 time out of 100 with this size sample if the workshop had no real effect on child-rearing skills. Thus, we can reject the null hypothesis and, if all other competing explanations for the apparent relationship can be ruled out, we can claim support for the research hypothesis.

TABLE 10.2 Child rearing skills scores for pregnant teenagers before and after the child rearing skills training workshop (*N* = 22)

Pretest	Posttest	Difference
4.8	5.8	1

$t = 2.53$, $df = 21$, $p < .01$

Independent Groups *t* Test

A third variation of the *t* test, the *independent groups* t *test,* is especially useful for research studies using a relatively small number of cases. Like the correlated groups *t* test, it also compares the means of two samples. But to use the independent groups *t* test correctly, the two samples must be independent of each other. That is, no case in one sample is connected to any case in the other sample, as they are when the correlated groups *t* test is used. The two groups do not have to contain an equal number of cases, which often is the situation in social work evaluations of treatment effectiveness.

Often two subsamples (e.g., clients in group treatment and clients in individual treatment) naturally tend to be unequal in size. Even in the most carefully designed research studies that randomly assign an equal number of cases to two subsamples, clients are likely to drop out of treatment before the study is completed. Any resulting discrepancy between the sizes of the two subsamples presents no problem for the independent groups *t* test. The formula for the test automatically controls for it.

When using the independent groups *t* test, cases selected for study are divided into two samples for the two-category independent variable (e.g., Intervention A/Intervention B). In experimental research designs, the two samples usually are the experimental group (Sample 1) and the control or comparison group (Sample 2). Mean scores of the two samples are compared using the formula for the independent groups *t* test, producing a *t* value and a corresponding *p* value. On the basis of chance, the means of the two research samples are likely to be somewhat different; the results produced by the test are an analysis of the amount of that difference.

If the difference turns out to be small given the degrees of freedom (the total number of cases in both samples minus 2 for this *t* test), so small that chance is a likely explanation for it, the null hypothesis cannot be rejected. We can conclude that the difference between the means of the two groups very likely is a function of chance and does not reflect a real relationship between the independent and dependent variables.

If the results indicate that the difference is large enough that it is unlikely to be the work of chance, we may be able to reject the null hypothesis and conclude that the difference observed in the sample indeed reflects a real relationship between the two variables. Chance will have been effectively discounted as one possible explanation for the apparent relationship between them. We must remember, however, that there still are three other alternative explanations that could have resulted from the study's research design: rival hypotheses, design bias, and sampling bias.

If a research sample is small, even a fairly large difference between two means may be due to chance. But there comes a point where a difference between two means is sufficiently large that chance alone is unlikely to have produced it. When is this point reached? The independent groups *t* test tells us. It determines the statistical likelihood of making a Type I error if we were to reject the null hypothesis and conclude that the difference between the two means is related to the presence of different values of the second (nominal) variable.

We will now return to an example used in Chapter 2. This time we will use it to illustrate the logic of the independent groups *t* test. Recall that we mentioned a hypothetical study guide that was developed to help social workers prepare for a state merit exam. To evaluate the effectiveness of such a study guide, we might randomly select 15 of the 30 social workers who plan to take the exam and provide them with a copy of it. We could give them specific directions to spend part of their study time each night using the guide as instructed.

The 15 social workers who used the guide could be regarded somewhat loosely as the experimental group; the remaining 15, who did not use it, could be regarded as the control group. After all 30 social workers took the state merit exam, their results could then be compared. We would not directly compare the individual scores of everyone who used the guide with the scores of everyone who did not. Instead, we would compare the mean exam score of the 15 social workers in the experimental group with the mean exam score of the 15 social workers in the control group.

In comparing the means for the two groups, we would need to ask certain questions. Is the difference in the direction that we predicted that it would be? Is it large enough to allow us to gain support to reject the null hypothesis (that there is no relationship between using/not using the guide and the social workers' scores on the exam)? How confident can we be that the difference was not due to chance? To put it another way, does the difference between the two mean scores probably reflect a real relationship between the independent and dependent variables?

Of course, as with all statistical testing, even if it can be demonstrated that a statistically significant relationship between the two variables exists, there are still those other possible explanations for it. Perhaps, for example, the guide served to remind the social workers to prepare for the exam. Consequently, they may have put more time and energy into studying than did those in the control group. Maybe the guide itself was of no direct help at all.

Even if all three possible alternative explanations could be ruled out, in addition to chance, another question still would have to be addressed: Is the relationship between the two variables a meaningful one? For example, does the size of the mean difference in scores on the exam justify the purchase price of the guide? The answer to this question, like others related to the practical value of the research findings, might be a difficult one.

As with the chi-square test, the popularity of the independent groups *t* test sometimes can lead to its misuse. The test is one that is familiar to us, easily understood, and relatively unintimidating to readers of research reports. Consequently, there may be a tendency to want to use the test, even in situations in which it is inappropriate and in which other, more appropriate tests should be used. Two common misuses of the independent groups *t* test are: (a) ignoring the shape of the distribution of the interval or ratio level dependent variable within the population and (b) using the "shotgun" approach to data analyses.

The independent groups *t* test (like all tests discussed in this chapter) is a parametric test (see Chapter 6). This means that it is designed for use in those data analysis situations when the interval or ratio level dependent variable is considered to be normally distributed within the population. If a frequency distribution for values of the variable within the population is noticeably skewed, other tests, such as one of the nonparametric alternatives presented in Chapter 11, should be used in its place. The value of research findings can be seriously jeopardized if an independent group's *t* test is used with interval or ratio level data that are not normally distributed within the population.

A second common misuse of the independent groups *t* test involves calculating such tests using a single dependent variable and a long list of dichotomous independent variables. In some particularly glaring examples of this "shotgun" procedure, researchers have run hundreds of tests with little basis in the literature for believing that the independent and dependent variables used might be related. To their delight, they have found a statistically significant relationship between one or two independent variables and the dependent variable.

Findings of one or a few statistically significant relationships between variables that are derived through using large numbers of independent groups *t* tests are not surprising. In fact, we would probably be more surprised if none of the relationships between the variables proved to be statistically significant! Probability theory alone would suggest that a "relationship" thus found may not be a real one. The "finding" probably is the work of a process sometimes referred to as "data dredging." It likely was produced by a variation on the old principle that "with an infinite number of monkeys, an infinite number of typewriters, and an infinite amount of time, some monkey, some time, somewhere, will write a Pulitzer Prize–winning novel!"

Similarly, if we try enough different combinations of variables, one or a few of them almost always will reflect statistical significance, at least the first time that we try them. Making too much of this can result in a Type I error, since the results are unlikely to be replicated. In situations where there is reason to believe that many different independent variables may be related to the dependent variable, other statistical tests should be used that are specifically designed for such situations.

The independent groups *t* test is the most commonly used of the three *t* tests among social work researchers. Both our micro and macro examples will demonstrate how it can be used.

Micro Example

Background.   Rose is a social worker in a large family service agency. In her agency orientation she was taught that the best format for marital counseling is to see both partners (husband and wife) together. Five years ago, she treated 20 couples, all of whom could only be seen individually (husband or wife) because of their work schedules. She was surprised to observe that, although they were never seen as a couple after their initial interviews, all 20 couples seemed to make excellent progress in solving their marital problems.

Over the years, Rose saw more and more couples on an individual basis. Since she believed she was having good client outcomes, she encouraged six of her colleagues to also counsel couples with marital difficulties by seeing them separately rather than together. The other social workers were also pleased with their clients' excellent progress.

Rose was not ready to conclude that the individual counseling format was really preferable to couple counseling. She decided to conduct a small-scale research study to see if she could find statistical support for her hunch that couples seen individually make better progress toward solving their marital problems than couples seen together.

Hypothesis.   As Rose began to search the social work literature, she found considerable support for the position that marital satisfaction is best enhanced when couples are treated in counseling together, not individually. But as she ventured into the literature from other fields such as psychology and pastoral counseling, Rose found a fair amount of support for the belief that success in marital counseling may be more likely to result from individual counseling. A reason sometimes given is that clients tend to discuss areas of dissatisfaction more readily and candidly when the spouse is not present.

Rose concluded that the literature was conflicting. However, she felt that her own observations and those of her colleagues were sufficient to tilt the balance enough to justify a one-tailed hypothesis:

Among couples receiving marital counseling, those seen individually will reflect a statistically significant higher level of marital satisfaction after ten weeks than those seen together.

Methodology.   Rose designed a small-scale research study to test her one-tailed hypothesis. She received permission from the agency director and her clients to randomly assign new clients who requested marital counseling during a three-month period to either individual or couple counseling. In a research sense, the clients were randomly assigned to one of two groups. All six social workers who had previously used and were experienced with both the couples counseling and the individual counseling formats participated as counselors in Rose's study.

Beginning the next month every other couple seen at intake was assigned to one of the six social workers to be seen together for counseling 50 minutes per week; the remaining couples were assigned to be seen individually for 25 minutes each per week. Those who could not agree to this arrangement were also seen but were not included as research participants in the study.

The *counseling method* (individual or couples) was the independent variable. The dependent variable *marital satisfaction* (to be measured after ten consecutive weeks of counseling) was measured using a widely used, standardized measuring instrument that measures marital satisfaction. It is believed to yield interval level data.

Fourteen couples were assigned to individual counseling sessions, and 14 were assigned to be seen as couples. Twelve couples in individual counseling completed ten weeks of treatment, and 13 couples in couples counseling completed ten weeks of treatment. After all 25 couples completed the marital satisfaction instrument, Rose took the score for each partner and averaged it with that of his or her spouse to get a "couple score." Then she computed the mean couple score of those who were seen individually (experimental group) and compared it with the mean couple score of those who were seen together (control group). The variable *marital satisfaction* as measured by the scale has been found to be normally distributed, so Rose felt justified in using the independent groups *t* test for her statistical analysis. She was attempting to determine whether the difference between the mean scores for the two groups was sufficiently large to allow her to reject the null hypothesis. She hoped to be able to conclude that a real relationship between the two variables was the likely explanation for the differences observed in the sample.

Findings.   The *t* value for Rose's data was 1.312. From a table of critical values of *t* (Appendix D), she learned that she needed a minimum *t* value of 1.714 to be able to reject the null hypothesis (using a rejection level of .05, a sample of 25 with 23 degrees of freedom, and a one-tailed hypothesis).

Conclusions.   Rose noted that if she were to reject the null hypothesis based on the analysis of her data, she would have slightly more than a 1 in 10 (10 percent) chance of committing a Type I error (1.312 is slightly smaller than 1.319). She clearly lacked statistical support for her one-tailed research hypothesis. Her initial disappointment was made even worse when she looked at the mean scores of the two groups. The clients who had participated in individual counseling scored somewhat worse on the average than did those seen together. Rose studied her findings some more. It was then that she realized that her lack of demonstrated support for a relationship between counseling method and marital satisfaction might reflect a useful finding in and of itself. Her inability to reject the null hypothesis could be interpreted to mean that it makes little difference which counseling method is used!

Rose also wondered how she could have been mistaken. The findings from her study were inconsistent with her previous impressions. She wondered whether she and the other social workers had perhaps merely perceived their individual counseling clients as doing better because of their surprise that these clients did *about* as well as those seen in couple counseling. Of course, she also wondered whether her hypothesis might still be correct. Perhaps the true relationship between the independent and dependent variables had been hidden by biased measurements or the influence of rival hypotheses (e.g., the social workers' greater experience with couple counseling). As she thought about it, Rose concluded that additional studies employing tighter research designs were indicated.

Implications.   In the interim, before further research studies could be conducted, Rose wondered what practical use she could make of her findings. During the next agency staff meeting she presented her study results. She was able to draw implications for social work practice within the agency. As frequently happens in social work research, her study generated more questions than answers. However, these questions served to focus the staff's attention on potentially productive areas of inquiry. Based on her findings, Rose and other staff members began to ask the following questions.

1. Since type of counseling (individual or couple) may have little or no effect in enhancing marital satisfaction, should continued attempts be made to encourage clients to enter couple counseling if they resist or if it represents a scheduling difficulty for them?
2. Should involvement of both partners in counseling continue to be a requirement for counseling, or should this policy be changed?
3. Should funds be allocated for a staff development program to enhance the social workers' use of individual counseling in treatment of marital problems?
4. Should the staff develop a single treatment model that combines individual and couple counseling, or should the professional staff be allowed freedom to select the counseling format that they prefer to use?

These questions and others ranged from issues that affected the individual social work practitioner to those that related to agency policies. The principal value of Rose's study was to call into question certain unchallenged practices within the agency and to encourage the staff either to justify or to discard them based upon further examination. Even if no changes resulted, the social workers would be practicing on a sounder theoretical base until subsequent research findings provided a more definitive answer to the questions.

Macro Example

Background.   Antonia is the director of social services for a large state health agency. In her professional role, she oversees social work services offered in the 50 district health offices across the state. It was recently brought to her attention that the agency was having a serious problem with social work staff turnover. A preliminary examination of agency data revealed that the problem was statewide and appeared to be normally distributed among the 50 district offices.

Antonia spoke with the personnel officer who was responsible for conducting exit interviews with employees leaving the agency. At first he preferred not to suggest possible reasons why so many social workers were resigning. But after Antonia assured him that she did not plan to ask him to identify workers who made complaints to him, he volunteered that the reasons given by many of the social workers for leaving appeared to be amazingly similar. He recalled that "many" of them seemed totally frustrated with their lack of autonomy in decision making. While the social workers recognized that in some professional and administrative matters the final decision had to be made by their supervisors, they saw no reason why many other decisions could not be made by them and their fellow professionals through a more democratic process.

Afterward Antonia thought about what she had been told. While her first inclination had been to be annoyed with the district supervisors for their apparent autocratic approach to decision making, she quickly realized that she had to take much of the responsibility for their supervisory style. She had come to rely on the use of staff authority when it seemed so effective with workers in her secretarial pool. She had hired trainers from outside the agency and required that all district supervisors attend training in the use of staff authority. She had also commented regularly at supervisors' meetings how effective she thought the approach was. Apparently the supervisors were only responding to Antonia's message that extensive use of staff authority is a sign of good supervision. By allowing their social workers only to advise them but not to make or implement decisions, the supervisors were really complying with Antonia's implicit directive.

Hypothesis.   Antonia knew that she needed an objective way to determine whether social workers' level of autonomy in decision making was related to staff turnover. She did not want to trust the impressions of the personnel officer without further data. She would not attempt to help all her supervisors to become more democratic in delegating decision making until she could be reasonably certain that some relationship existed between the two variables. She decided to conduct a small research study to test the one-tailed hypothesis:

> There will be a statistically significant, lower rate of staff turnover in democratic decision-making environments than in autocratic ones.

Methodology.   Antonia knew that recent management literature stressed the use of quality circles as a promising way to solve some administrative problems by arriving at decisions via the group process. (The approach is characteristic of Theory Z management methods that have been viewed as successful in Japan for many years.) Antonia had been considering the use of quality circles anyway and saw this as a good time to try them. Quality circles seemed to her to be a good way to create a more democratic approach to decision making in the district offices.

Antonia randomly selected ten districts to serve as her experimental group. She then provided release time to the ten district supervisors to attend an out-of-state workshop on the use of quality circles. She told the supervisors that she expected them to implement quality circles in their supervision, and she requested a report on their methods of implementation to ensure that this had been done. They were asked not to share their experiences with other supervisors.

At the same time, Antonia randomly selected ten other districts as her control group. Their district supervisors were given no additional training and no new instructions on how they should handle decision making in their work groups.

After one year Antonia computed the average rate of turnover for districts in the experimental group and for those in the control group. She had two categories of the dichotomous nominal independent variable *decision-making environment,* democratic and autocratic. Her dependent variable *turnover rate* was at the ratio level of measurement. The situation seemed to be well suited to the use of the independent groups *t* test to test whether there might be statistical support for her one-tailed hypothesis.

Findings.   Antonia compared the average turnover rate for the two groups using the independent groups *t* test. The average in the experimental group was lower than in the control group. The *t* value for Antonia's data was 1.992. She concluded that there were 18 degrees of freedom (10 + 10 = 20 – 2 = 18). Using a table such as Appendix D, she noted that, for the row corresponding to 18 degrees of freedom, the *t* value from her statistical analysis fell between 1.734 and 2.101. She moved to the left (1.734) and observed that it was in the column headed by .05 for one-tailed tests and .10 for two-tailed tests. The *p* value corresponding to her data under the one-tailed hypothesis was therefore less than .05.

Conclusions.   From her knowledge of statistics, Antonia knew that she had found support for a statistically significant relationship between the independent and dependent variables. She knew that, if she were to reject the null hypothesis, she would have less than a 5 percent chance of committing a Type I error on the basis of statistical probability alone. She also was pleased to see that the relationship was in the direction predicted; the mean turnover rate for district offices in the experimental group was lower than the mean turnover rate for district offices in the control group.

Because her study had been very limited in scope, Antonia was reluctant to view her findings as an unequivocal endorsement of the expanded use of quality circles or other more democratic methods of decision making as a way to reduce social work staff turnover. She recognized that it would be precipitous on her part to proceed to implement her findings as though she had uncovered a simple cause-effect relationship. The weak research design of her study had certainly not eliminated the three alternative explanations (rival hypotheses, design bias, and sampling bias) as possible explanations for the difference in staff turnover rate between the two groups of district offices.

Certain methodological questions remained. For example, how much did the opportunity to go out of state for training positively affect the morale of the experimental group supervisors? Perhaps they came back in a better mood and more inclined to be considerate of their workers. If so, this may have been a better explanation than the implementation of quality circles for their lower staff turnover rate. Or did the workers view the supervisors in the experimental group as more considerate simply because they made an effort to try something new? If so, this might have been a major factor in reduced turnover rate.

Implications.   Despite the fact that Antonia's findings would have to be viewed as tentative, she was still able to use them both to understand the problem of staff turnover and to begin to address it through her actions. In light of her findings, she asked herself these questions:

1. How can I adjust my supervisory style with district supervisors so that I do not unintentionally communicate to them that I expect extensive use of staff authority relationships in their supervision of social workers?
2. How can I help district supervisors to identify decisions that are more appropriately made using the democratic process? How can I make district supervisors feel more comfortable using that process?
3. How can I help district supervisors to identify decisions that are inappropriate for the democratic process (e.g., personnel matters) and to continue to use staff authority and other less democratic approaches for these decisions without harming the morale of workers?
4. What use of the democratic process, besides quality circles, would help social workers feel that they have more input into decisions if they possess the necessary expertise?
5. Would it be advisable to send all district supervisors to quality circle training out of state?
6. What further research studies can be designed to provide additional support for the finding that democratic decision making is associated with lower staff turnover rates?

After thinking about these and other questions that emerged from her study, Antonia discussed her ideas for implementing her findings with friends who are social work administrators in other large agencies. She then settled on a plan of action.

At the next meeting of all the supervisors, Antonia shared the findings of her research study. She reiterated her support for the use of staff authority in certain situations but also stated her belief that social workers at all levels are professionals and need to be involved in decision making. She emphasized that she believed overreliance on autocratic supervisory approaches can hurt morale and, even more important, does not take advantage of the expertise of other staff in addressing problems. She supported these contentions by asking a former member of the experimental group to use part of the next supervisors' meeting to teach all the supervisors (including herself) the basic principles of quality circles. In the meeting, the other nine supervisors who had used them were asked to share their experiences as well.

Antonia was convinced that sharing decision making in certain situations by district supervisors with their staffs was indicated. In another meeting she emphasized this belief to the supervisors. Consistent with it, she allowed them to decide individually whether they preferred to receive training in quality circles so as to implement that technique in their districts or to develop their own plans (with her approval) for introducing ways to increase democratic decision making among their staffs.

Finally, Antonia set aside time to develop a more comprehensive and rigorous analysis of the problem of staff turnover. The study would examine other factors, in addition to approaches to decision making, that the literature suggests are related in some way to staff morale and turnover. She hoped that ultimately the findings from a larger-scale study could be generalized to other social work settings and that a report of the findings would have the potential for publication in a professional social work journal.

ANALYSIS OF VARIANCE (*ANOVA*)

Another group of related statistical tests is referred to as *analysis of variance* (*ANOVA*). These tests are appropriate for use in hypothesis-testing situations when *t* tests cannot be used. They are useful when the means of three or more groups are compared. *ANOVA* is commonly used when the dependent variable is at the interval or ratio level of measurement and the independent variable has three or more categories. It should be apparent by now that a *t* test can be used when there are only two groups. While *t* tests produce a *t* value that can be compared with a table to determine whether or not it is justifiable to reject the null hypothesis, the variations of *ANOVA* produce an *F* ratio (with its own table) that allows us to make the same determination.

One-Way *ANOVA*

Sometimes we have only one interval or ratio level dependent variable and three or more different values of the independent variable. The independent variable may be either nominal or ordinal, precluding the use of Pearson's *r*, which, we will recall, is designed for use with two interval or ratio level variables. However, we still want to use a relatively powerful parametric test to examine a possible relationship between the two variables.

For example, we may be using three or more counseling approaches that are designed to affect assertiveness of clients. We want to know if one of these approaches really is more effective than the other two. Or we may want to examine whether clients categorized at intake as high, medium, or low for the variable *treatment motivation* differ significantly in their length of psychiatric hospital stay.

With one-way *ANOVA*, we can determine if the differences in the means of three or more subsamples for the interval or ratio level variable are great enough that chance is likely to have caused them. Suppose that, for example, we wish to determine if there is support for the two-tailed research hypothesis that, "Among treatment staff in a psychiatric setting, professional discipline of the therapist is related to client satisfaction with treatment." The null hypothesis is, "Among clients seen by therapists of different professional disciplines, there is no real difference in client satisfaction level."

We could use a standardized scale that yields interval or ratio level data to measure what logically would be the dependent variable *treatment satisfaction.* However, since we have four categories of the independent variable *professional discipline of therapist*—psychiatrist, psychologist, social worker, and psychiatric nurse—the use of six *t* tests would be inappropriate. We should not compare the six possible combinations of pairs of means using six independent groups *t* tests—that is, psychiatrist and psychologist, psychiatrist and social worker, psychiatrist and nurse, psychologist and social worker, psychologist and nurse, and social worker and nurse. As discussed previously, if this was done, we would be using the same means repeatedly and thus increasing the likelihood of chance producing a statistically significant relationship. One-way *ANOVA* is the better choice.

One-way *ANOVA* would look at the mean treatment satisfaction for each of the four subsamples, or categories (professional disciplines). Unlike the correlated groups *t* test, the subsamples used in *ANOVA* need not be of equal size. A *grand mean,* the mean treatment satisfaction for all clients in our study, would also be computed as part of the formula for one-way *ANOVA*. Very likely the mean of each subsample would differ from the others as well as from the grand mean. *ANOVA* would examine these differences (referred to as the *between groups variance*) as well as the average amount of variability of treatment satisfaction within each subsample (referred to as the *within groups variance.*) It also would look at all the treatment satisfaction scores and how they are distributed (vary) around the grand mean.

It should be pointed out that *ANOVA* does not tell us which subsample mean is statistically different from the others. It only tells us if there is or is not a statistically significant relationship among the four subsample means. If there is a statistically significant relationship among the subsamples, then we need to find out which subsample(s) is statistically different from the others. There are many ad hoc statistical methods that can help us with this, including one called the Duncan test. However, most statistical computer software packages now will do this analysis for us.

From the above description, it should be obvious that computing one-way *ANOVA* by hand would be very time consuming. A computer is a virtual requirement. The complexity of the formula precludes hand calculation, but it also is what makes one-way *ANOVA* a relatively powerful parametric test. It can identify relationships between variables that other, less powerful tests might miss. However, as a general rule, more powerful tests also are more restrictive than less powerful ones—that is, more conditions for their use must be met. *ANOVA* is no exception.

For example, in addition to the requirements of the other parametric tests, one-way *ANOVA* should be used only when the amount of variance of scores within each subsample is approximately equal. In our hypothetical study of the relationship between professional discipline of therapist and treatment satisfaction, we should not use one-way *ANOVA* if, for example, the amount of variation in satisfaction among clients seen by psychiatrists tends to be much greater than among clients seen by one or more of the other professionals.

Factorial Designs

Other variations of *ANOVA,* called *factorial designs,* are available for situations in which we wish to examine the relationship among two or more nominal or ordinal level independent variables and one interval or ratio level dependent variable. For example, we could use two-factor ANOVA to examine the relationship among the discipline of the therapist and the gender of therapist and the dependent variable *treatment satisfaction.* The relationship would be a complex one, and the formula is correspondingly complex since we are now venturing once again into the area of multivariate analysis—statistical testing that has the potential to examine the interaction among three or more variables.

Three-factor *ANOVA* is used as we might guess it is used. It examines the relationship among *three* nominal or ordinal level independent variables and one interval or ratio level dependent variable. Still another type of test altogether, Hotelling's T^2, is appropriate when there are two or more dependent variables as well as two or more independent variables. A discussion of the complexities of this test and of the special uses and requirements of *ANOVA*'s factorial designs is outside the scope of this book. An advanced statistics text is recommended for those readers who wish to go beyond our brief mention of multivariate statistics in this chapter and elsewhere and our overview of the topic in Chapter 12.

CONCLUDING THOUGHTS

The family of related tests that involve a comparison of two or more means are well suited to social work research. They are both versatile and powerful. *T* tests (for use with two means) and *ANOVA* (for use with three or more means) are being used increasingly for social work hypothesis testing.

The examples in this chapter have illustrated how *t* tests and *ANOVA* can be used by social work researchers and practitioners. While these tests often are used in major research projects that have extensive funding and use sophisticated research designs, they also are valuable for preliminary, limited efforts. In addition, while those statistical findings that support hypotheses are of practical value at many different levels for the social work practitioner, nonsupport for research hypotheses can be of equal value. Conscientious researchers cannot lose. If a study is designed and implemented well and statistical testing is conducted correctly, we can advance the body of knowledge available to social work practitioners, whether statistical support for a research hypothesis is found or not.

STUDY QUESTIONS

1. What is the appropriate combination of levels of measurement of two variables for the use of a *t* test?

2. Why do the sample size and subsample size comparability requirements of the independent groups *t* test frequently make it ideally suited for social work research?

3. If, using the independent groups *t* test, the null hypothesis were correct, would the mean value of a variable in one sample be very similar to or very different from the mean value for that variable in the other sample?

4. What do *t* requirements say about using *t* tests with interval or ratio level variables that reflect a skewed distribution within the population?

5. What are the formulas for degrees of freedom for the three *t* tests described in this chapter?

6. What additional step is required in determining whether a *t* value that is statistically significant reflects support for a one-tailed hypothesis?

7. Explain why a *t* test that does not result in a finding of statistical significance may still produce a finding that is useful for the social work practitioner. Provide an example.

8. Provide an original example of how one could use a one-sample *t* test to evaluate practice effectiveness within a social agency with which you are familiar.

9. Find an article in a professional social work journal that reports on the use of the correlated groups *t* test. What was the study's hypothesis? How did the author interpret the study's results? Do you feel the test was used appropriately? If so, why? If not, why not? What contribution did the article make to the profession's knowledge base?

10. Find an article in a professional social work journal that reports on the use of the independent groups *t* test. What was the study's hypothesis? How did the author interpret the study's results? Do you feel the test was used appropriately? If so, why? If not, why not? What contribution did the article make to the profession's knowledge base?

11. Find an article in a professional social work journal that reports on the use of one-way *ANOVA*. What was the study's hypothesis? How did the author interpret the study's results? Do you feel the test was used appropriately? If so, why? If not, why not? What contribution did the article make to the profession's knowledge base?

chapter 11

Nonparametric Options

The preceding three chapters presented some of the most widely used statistical tests that can be used within social work research studies—chi-square, Pearson's r, and tests that compare two or more means (t tests and $ANOVA$). Although they are used to analyze data in a large number of research situations, there are many data analysis situations where they cannot be used. For example, as previously mentioned, certain requirements for expected frequency size must be met in order to use the chi-square test. Also, if interval or ratio level data are not available for at least one variable, and/or are not normally distributed within the population, then parametric tests cannot be used.

Fortunately, there are good nonparametric alternative tests available. Over the years, a number of well-known (and not so well-known) nonparametric tests have proven to be effective tools in the analysis of social work research data. The objective of the brief discussion that follows is to provide enough information about several nonparametric tests to enable readers to gain some general insight into the way the tests work and the kind of social work research and practice situations in which they might be used. Once having determined that a test may be appropriate, it would then be necessary to consult an advanced statistics book to make the final decision as to the test's appropriateness and to gain the additional details about its use.

OPTIONS FOR USE WITH NOMINAL MEASUREMENT

When only nominal level data are available and the requirements for the use of the chi-square test cannot be met, two good nonparametric alternatives can be used: the McNemar's test and the Fisher's exact test.

McNemar's Test

The *McNemar's test,* also referred to as the test for the significance of changes, is popular in research situations that employ a one-group pretest-posttest research design that entails the measurement of a two-category nominal level variable twice within a single research sample. Thus, the McNemar's test is frequently used in determining if a type of social work intervention may have had an impact. For example, it could be used to determine if an educational program on the life experiences of Southeast Asian refugees seemed to produce a desired change in attitudes about them among those who attended the program.

In social work practice and research we often are interested in affecting a dependent variable in some desired direction. Often, we may wonder whether an intervention is having any effect at all. Because of the existence of so many other variables that can influence attitudes or behaviors, it may be hard to know just how influential our intervention may have been in any changes in the dependent variable that occurred. The McNemar's test offers the opportunity for at least a preliminary insight into these issues. It can tell us whether an intervention is associated with change at all, and it can help us to determine if the changes that occurred were in the desired direction.

Example.   Chester is a school social worker. He requested time to address a meeting of a parents' group to present his arguments for hiring four additional school social workers for the district, a proposal currently being considered by the school board. He wondered whether the parents really would be influenced by his arguments one way or the other or would just listen politely. Before Chester even considered giving up more of his evenings to speak to other parents' groups in the hope that they would try to influence school board members, he needed to know if his first presentation would be associated with a change in parents' thinking (the dependent variable) and, if so, if the change would be in a desired direction.

Chester speculated that his presentation (the independent variable) would influence parents' thinking on the issue. But he was not sure in which direction they would be influenced. Thus, he concluded that he had a two-tailed hypothesis. He predicted that change would occur, but he did not predict the direction of the change.

Following his presentation to the group, Chester gave a sheet of paper to each parent. On it he asked them to indicate if (a) if they had been in favor of hiring the new social workers before his presentation (pretest) and (b) if they were in favor of hiring them following it (posttest). Based on the parents' answers, each parent fell into one of four categories, or cells:

1. Favored before, favored after—cell *a*
2. Favored before, did not favor after—cell *b*
3. Did not favor before, favored after—cell *c*
4. Did not favor before, did not favor after—cell *d*

TABLE 11.1 Significance of change analysis (McNemar's test): Positions on proposal before and after Chester's presentation

Before Presentation	After Presentation		Total
	For	Against	
For	6 (a)	26 (b)	32
Against	10 (c)	8 (d)	18
Totals	16	34	50

$x^2 = 5.58$, $df = 1$, $p < .02$

Chester took his data and placed them into a 2-by-2 contingency table similar to one used with chi-square (Table 11.1). He then used the chi-square formula (with the Yates Correction) factor to compute the chi-square value. He checked the table of critical values on a table like Appendix B to see whether he had statistical support for rejection of the null hypothesis.

In Table 11.1, the totals on the right side represent the number of parents for and against the proposal before (pretest) Chester's presentation (32 and 18, respectively). The totals along the bottom of the table represent the number for and against it after (posttest) the presentation (16 and 34, respectively). The numbers within the body of the table (its four cells) represent individual cases, as they do in a cross-tabulation table where chi-square is used. The 50 cases (parents) were distributed among the table's four cells based on a pair of measurements—whether or not a parent favored the proposal before the presentation and whether or not a parent favored the proposal after it. Thus, if Ms. Aguilar favored the proposal before the presentation but opposed it after it, she would be one of the 26 cases in cell *b*. Or if Mr. Owens was against it before the presentation and was still against it after the presentation, he would be one of the 8 cases in cell *d*.

The McNemar's test looks at change. Cells *b* and *c* include individuals who changed their position; cells *a* and *d* contain cases who had *no change* in their position. Since the McNemar test is really not very interested in cases where no change occurred (cells *a* and *d*), the focus of its analysis is on cells *b* and *c*.

The null hypothesis would suggest that some change would likely have occurred with some cases, but that whatever change occurred probably would not be the result of Chester's presentation. Logic would state that about half the change that occurred would be in one direction and half in the other direction. These changes would theoretically "cancel each other out," and the numbers favoring or opposing the proposal to hire more social workers would remain about the same after the presentation as they were before it.

Statistically, the McNemar's test seeks to determine whether the null hypothesis can be rejected by demonstrating that the preponderance of change that occurred was in only one direction. From the data in Table 11.1, Chester was able to reject the null hypothesis ($p < .02$). Unfortunately, the direction of most of the changes that occurred indicated that his presentation may have been associated with negative results (parents' turning against the proposal) rather than positive ones (parents' deciding to support the proposal).

Only 14 parents failed to change their positions (cells *a* and *d*). But of the 36 who changed them (cells *b* and *c*), only 10 moved from negative to positive (cell *c*), while 26 (cell *b*) who had previously favored the proposal opposed it after the presentation. The demonstration of a statistically significant change was an endorsement of Chester's ability to influence parents' attitudes. However, the results of his presentation clearly were not what he had sought.

Chester speculated on the meaning of his research findings. Had he said something to "turn off" the parents? Why had so many of them been negatively affected by his presentation? Before he gave any more presentations to parents' groups, he planned to review carefully what he had said and how he had said it.

As can be seen, the McNemar's test is quite limited. Usually it is used in research situations that involve a two-category nominal level variable that is measured twice for the same sample of cases. Because it is ideal for many one-group pretest-posttest studies designed to evaluate the effect of a method of practice intervention, it can be a useful test for the social work researcher or practitioner. It often is used as a preliminary to more high-powered statistical analyses.

Fisher's Exact Test

The *Fisher's exact test,* like chi-square (and unlike McNemar's test), is used with two *independent* subsamples. The independent subsamples may have resulted from the random sampling of cases that have been categorized as falling into two identifiable groups, such as being in either individual counseling or group counseling, gender (male/female), or marital status (married/not married). Or they may be the experimental and control groups to which cases were deliberately and randomly assigned in a classical experimental research design.

The principal advantage of the Fisher's exact test over chi-square is that it can be employed when the expected frequencies in a cross-tabulation table for the data are too small to meet the criteria for using chi-square (i.e., expected frequency in a given cell is less than 5). However, unlike chi-square, the test has one serious limitation—it can only analyze data using a 2-by-2 contingency table (four cells). If we had more than two values of either or both of the two variables that we think may be associated, it would be necessary to collapse the data into a 2-by-2 table before the Fisher's exact test could be used. This would result in a loss of available measurement precision, which is something that should be avoided if possible. If feasible, it might be preferable to enlarge the size of the research sample so that the requirements for the use of the chi-square statistic can be met.

Like the McNemar test and another simple test that requires only nominal level measurement, Yule's *Q*, Fisher's exact test often is used to obtain a preliminary answer to a research question. It is easily computed by hand. Subsequent to its use, more sophisticated analyses can be attempted using larger samples and/or more rigorous research methods.

OPTIONS FOR USE WITH ORDINAL MEASUREMENT

Previous chapters presented several different parametric statistical tests that require interval or ratio level variables that must be normally distributed. What can we do if these two requirements are not met? There are several good nonparametric alternative tests that can be used when data are only at the ordinal level of measurement and/or they tend not to be normally distributed. Below are six of the most popular ones.

Spearman's Rho and Kendall's Tau Tests

Two nonparametric tests that require only ordinal level data are commonly used in situations where, if normally distributed interval or ratio level data were present, Pearson's *r* would be used. They are discussed together because they are so similar that they may be considered almost interchangeable. They are Spearman's rho and Kendall's tau.

Like Pearson's *r*, both *Spearman's rho* and *Kendall's tau* produce a correlation coefficient that is either positive or negative and that has a numerical value between +1.0 and –1.0. But while Pearson's *r* uses all the case values of two variables in its formula to see to what degree and in which direction (positive or negative) the variables are correlated, both of the nonparametric alternatives focus on the *ranks* of research participants (or objects) for the two variables (rather than their exact scores on them). For example, their formulas would use the fact that a given person might have ranked highest in the research sample on the ordinal variable *level of motivation* and third on the variable *socioeconomic status*. Another individual may have ranked fifth on both variables, another second on the first and seventh on the latter, and so on.

The actual case values for each pair of variables, while useful in determining an individual's rankings on them within the sample, are not used in the formula for either Spearman's rho or Kendall's tau. This makes the tests about equally powerful (i.e., able to detect the presence of a real correlation between two variables) but slightly less powerful than Pearson's *r*. When Spearman's rho and Kendall's tau are both computed using case rankings for two variables for the same data set, they generally produce slightly different correlation coefficients but very similar *p* values.

Like most statistics, both Spearman's rho and Kendall's tau are available on many personal computer statistical software packages. They can, of course, be computed by hand as well. However, if this must be done, two factors must be considered in determining which of the several slightly different formulas should be used: (a) the size of the sample and (b) the number of "ties" (more than one case sharing a ranking for a variable) present within the raw data set.

Example. Meredith is a social worker in a hospice that provides a variety of services to caregivers of terminally ill patients. After working with hundreds of patients and their caregivers, it seemed to her that cancer patients whose caregivers were initially willing to use hospice services were more likely to live beyond their projected date of death than those who were more reluctant to use them. Meredith decided to test the following one-tailed hypothesis: "There is a positive correlation between caregivers' willingness to receive hospice services and the longevity of the patients." She tested this hypothesis by conducting a research study using the nine patients who signed up for hospice services during the next month.

At the time that they signed up, she and another social worker, who sat in on the initial interview, discussed each caregiver and then categorized them as "very reluctant," "somewhat reluctant," or "eager" to receive services. Then she followed each patient and his or her family until the patient's death, and she noted for each whether the patient died "before," "at about," or "after" the time of death the patient's doctor had projected when services were initiated.

Meredith took her data on the nine patients and examined the relationship between the predictor variable *willingness to use services* and the criterion variable *longevity of the patient* using the Spearman's rho statistic. She noted that her sample was small and that there were many ties in the rankings that she compiled for the two variables, so she used the appropriate formula for these conditions. It produced a correlation coefficient (r) of .2461 and a corresponding p value greater than .05. By glancing at Appendix C, it can be easily seen that she needed a minimum correlation coefficient of .5822 for her data to be statistically significant at the .05 level (with only nine cases).

Based upon her statistical analysis, Meredith concluded that she could not reject the null hypothesis and she could not claim support for her research hypothesis. She concluded that her observations may have been a bit of "wishful thinking" on her part. However, just in case her hypothesis had been correct, she decided to test it again using a larger number of cases active in her agency over a one-year period.

Since Spearman's rho and Kendall's tau are both tests of correlation, they (like Pearson's r) are used to determine the strength and direction of a relationship between variables. Had Meredith found support for her hypothesis, the finding might also have been helpful in predicting the longevity of patients based upon their caregivers' attitudes toward their use of hospice services. It would not have suggested that the predictor variable caused the variations in the criterion variable.

The two tests bear other similarities to Pearson's r as well. Chapter 8 briefly described *partial r*, a variation of the Pearson's r, which controls for the effects of a second predictor variable. A similar nonparametric variation of Kendall's tau is available for use with ordinal data—*Kendall's partial rank correlation coefficient*. Like partial r, it is most useful in situations where there is one potentially intervening or extraneous variable that cannot be controlled for by the research design and that represents an obvious rival hypothesis for explaining the variations in values of the criterion variable.

Mann-Whitney U Test

Probably the most frequently used nonparametric analog to the parametric independent groups t test discussed in Chapter 10 is the *Mann-Whitney* U *test*. It is used with a dichotomous nominal level independent variable and an ordinal, interval, or ratio level dependent variable that is not normally distributed (as required for use of the independent groups t test). Like the independent groups t test, the U test attempts to reject the null hypothesis that states that the two subsamples are not very different with respect to their measurements of the ordinal or interval level dependent variable.

The U test is especially useful in research studies involving two small independent subsamples (the dichotomous independent variable). It is frequently used in quasi-experimental situations to determine whether a "treatment" given to the experimental group, but not to the control group, appears to result in a difference in regard to some dependent variable that can be rank ordered (ordinal). Like the independent groups t test, the U test does not need two groups of identical size. The same formula is used with very small samples (under eight) or larger ones, but we must be careful in interpreting results to use the table of critical values that is designed to adjust for the size of the sample used.

The U test is easily computed with a pocket calculator when the sample is small. It is based on the assumption that a good indicator of difference between the two groups is the number of cases in one group that fall below each respective score of the other group when all scores are rank ordered. The logical premise underlying it is that the presence of a disproportionate number of higher scores drawn from one group and of lower scores drawn from the other group probably suggests that, on the whole, the two groups really are different in regard to the dependent variable. The U test is a mathematical way of determining whether this pattern is sufficiently strong to be able to reject safely the null hypothesis.

Example. Shanti is a social worker employed by a large corporation. During July he had 17 referrals of workers identified as having an "attitude problem" on their job. He asked each of their supervisors to name another worker at the same level whom they would describe as having a "very good attitude" who might participate with him when the problem worker was being counseled. Seven of the supervisors complied with his request; the other ten did not.

Shanti saw all 17 workers (the seven with his or her co-worker and the other ten alone) for five sessions. After the fifth interview, he asked their respective supervisors to complete a job attitude scale that had recently been developed in the personnel department for supervisors to evaluate the corporation's workers. He then rank ordered the scores of all 17 workers and used the U test to test for group independence. He was optimistic that the "experimental" counseling approach involving a co-worker would be found to be superior to the supervisor-supervisee approach used in the control group. He quickly noted that three of the top four attitude scores were achieved by experimental group members.

In fact, the U test did not allow Shanti to reject the null hypothesis. Two of the lowest scores also came from experimental group members. Even if the U test had achieved statistical significance, alternative explanations to a true relationship (lack of random assignment to groups, the effect of other variables, and so forth) could not have been ruled out because of the lack of rigor of Shanti's research design. Shanti decided to tighten up the design and to pursue his inquiry further by using larger samples and by including a pretest of worker attitudes.

Like other nonparametric options that we have discussed, the U test often is used as a "preliminary method" of examining relationships between two variables without investing great amounts of effort in research design and data analyses. Because it requires only ordinal level data, it is appropriate for situations in which the development of a new data collection instrument precludes any claim by the research study that there is adequate measurement precision to generate normally distributed interval or ratio level data. Because it requires only a small sample size, it also lends itself well to a preliminary analysis of possible relationships.

As we have suggested, the U test is a useful test for comparing the rankings of a variable within one subsample with those within a second subsample. If there are more than two subsamples (values) of the dichotomous nominal independent variable, there is another, related nonparametric test that can be used. It is the *Kruskal-Wallis test*. It works very much like the U test, ranking all scores, sorting the rankings into subsamples, and then examining how many cases in other subsamples are ranked below each case in a given subsample. Generally speaking, Kruskal-Wallis is the nonparametric analog of the parametric one-way *ANOVA*. We could state this as an analogy: Kruskal-Wallis is to the U test what one-way *ANOVA* is to the independent groups t test.

Median Test

Another easily calculated nonparametric test for situations where we lack a normally distributed interval or ratio level dependent variable is the *median test*. Like the independent groups t test, it employs a comparison of a measurement of central tendency. But unlike the independent groups t test, which compares two means, the median test examines those cases that fall above and below the median for a dependent variable. Like other statistical tests, it is designed to assess the likelihood that two subsamples (sometimes an experimental group and a control group) are sufficiently different from each other in regard to the dependent variable to warrant rejection of the null hypothesis.

The median test is relatively simple. Scores are rank ordered for the ordinal or skewed interval level dependent variable, and the median for all scores is computed. Then the number of scores above and below the median (those at the median are simply dropped or added to one group or the other) are tabulated for both categories of the dichotomous nominal level independent variable, and the frequencies are placed into a 2-by-2 contingency table.

The median test divides the sample into two nearly equal or equal-sized groups (half above and half below the median) for the dependent variable. With larger samples (30 or 40 or more total cases) that also reflect a reasonably even split in the number of cases in the two values of the nominal level variable (subsamples), the expected frequencies may be large enough to justify the use of the chi-square test to complete the testing process. When the total number of cases (both groups) is small, Fisher's exact test is substituted to see if the relationship between the two variables is statistically significant.

The median test is based on the assumption that, if the two groups are not really different or reflect differences largely attributable to chance, each will have approximately the same percentage (subsample sizes need not be equal) of cases above and below the median. A clustering above the median by one group and below by the other may indicate a real relationship between the two variables, depending on the strength of this pattern and the degree to which other explanations have been ruled out.

Example.   Beulah is a social worker employed in a genetic counseling center. She observed that only about half of pregnant women over 40 years of age who were referred for amniocentesis (a test to determine if the fetus has any of several genetic problems) followed through with the referral. In her rather unscientific observation, she noticed what she thought might be a factor related to this phenomenon. It seemed to her that women who did not follow through on the referral tended to have several children already; those who did follow through seemed to have fewer children.

The variable *number of children* is at the ratio level of measurement. But Beulah knew that the variable was not normally distributed. She decided to use the median test. She grouped the data into four cells of a contingency table based on where each case fell in regard to the dependent variable *completed/uncompleted referral* and the independent variable *number of children* (above or below the median).

The criteria for size of expected frequencies for chi-square were met, so she completed her analysis using the appropriate formula. Her initial impressions were confirmed by her use of the median test. She speculated on her findings, wondering whether the possible birth of a child with, for instance, Down's syndrome was of less concern in larger families because additional child-care help was available from older siblings. She decided to pursue the possible explanation in subsequent research studies.

The example is a little unusual for use of the median test in that it was the independent variable that was treated as ordinal level data. (It was really ratio but badly skewed.) More typically, it is the dependent variable that is at the ordinal level, while the independent variable (e.g., *type of treatment*) is at the nominal level. The test is a good one when either combination of measurement precision is present.

The median test, unlike some of the others presented in this chapter, is included here because it is potentially useful to the social work researcher, not because it is commonly used in social work research. All too frequently, for example, we have observed that the criteria for the independent groups *t* test cannot be met, and we quickly have turned to chi-square.

In many instances, the median test is a preferable alternative. It treats the measurement of the interval or ratio level dependent variable as ordinal, thereby using more available precision of measurement than does chi-square, which treats all variables as if they are only at the nominal level of measurement. To use chi-square in such situations is to run the risk of throwing away available precision of measurement and to increase the risk of an error in drawing conclusions from the findings of a research study.

Kolmogorov-Smirnov Two-Sample Test

The *Kolmogorov-Smirnov test* (hereafter referred to as the *K-S* test) has some similarities with the median test. However, it compares more than just central tendency data. It compares (between two samples) the dispersion, skewness, and other characteristics of the distribution of values of an ordinal level dependent variable, that is, the entire shape of their distributions. (A variation, the *K-S* one-sample test, works similarly, except that it compares the distribution of a variable within a sample with another, theoretical distribution of the variable.)

The *K-S* test is based on the assumption that if two subsamples (dichotomous independent variable) are randomly drawn from the same population and the null hypothesis is correct, variations of the ordinal or skewed interval or ratio level dependent variable should be very similar for each of the two categories of the independent variable. If differences between the two categories are considerable, they are probably not just random deviations that exist because of chance but instead indicate a real relationship between the two variables. In short, the test is a way of determining if the differences in the distribution of the values of the ordinal or skewed interval or ratio level variable are large enough to rule out chance and to consider rejection of the null hypothesis.

If the highest level of measurement used is at the ordinal level, the *K-S* analyzes the data by comparing the cumulative frequencies for intervals in the variable (e.g., how many cases were rated "slightly improved" for the control group versus the experimental group). More specifically, the test focuses on the point (interval) at which the cumulative frequency difference between the two subsamples was the largest.

Example.   Roosevelt is a county director in a public assistance agency. Several of his African-American workers had complained that Caucasian workers seemed to be displaying excessively suspicious and punitive attitudes toward their AFDC clients. They stated that they believed that the clients who were being seen by Caucasian workers were not trusted and were assumed to be "cheaters."

No valid and reliable single measurement of suspicious and punitive attitudes was immediately available, so Roosevelt concluded that the number of referrals for fraud investigation was a beginning operationalization of the dependent variable. Although an exact count of referrals was possible, when used as a measurement of attitudes (the real dependent variable), ordinal measurement of the variable was all that could be claimed. He compiled a count of fraud referrals for each worker during the previous three months.

Roosevelt then assembled a cumulative frequency distribution for African-American workers and for Caucasian workers, using eight intervals of number of referrals for each. The decision to use eight intervals (referrals ranged from 9 to 32) was made in an effort to "conserve" the precision of measurement available (using three intervals would throw away too much data) while not "cutting too thin" by claiming that, for example, a difference of only one referral was a valid indicator of a real difference in attitude. The decision relied primarily on common sense and research ethics.

Having identified the interval at which the cumulative frequency distributions were most different from each other, Roosevelt applied the *K-S* formula. He learned that, in fact, there was little real difference between African-American and Caucasian workers for the measurement of the variable, at least not enough to rule out chance and to be able to reject the null hypothesis. While he did not totally discount the complaints of his African-American workers (a better measurement might have revealed an attitudinal difference), he did feel that he lacked sufficient evidence to confront his Caucasian workers about what some fellow workers perceived as a problem. He chose instead to reinforce what he believed to be appropriate attitudes toward AFDC clients in future staff meetings.

The *K-S* test is useful in looking at the relationship between an ordinal level variable and a nominal level dichotomous independent variable when just a comparison of central tendency may be insufficient. By examining the whole distribution of values of the ordinal level dependent variable, we get a more complete picture of the similarity of two subsamples. Still another test, the *Wald-Wolfowitz runs test*, goes even further in that it identifies subtle differences in distributions of variables that the *K-S* test does not detect.

Wilcoxon Sign Test

Social work researchers often find themselves with measurements that are a little more precise than ordinal but not quite precise enough to qualify as interval level. This is frequently the case with newly developed instruments that measure attitudes, perceptions, or beliefs. Often, just the amorphous nature of what is being measured prevents us from claiming that the value labels we assign reflect precise intervals—that is, equal difference in quantity of the variable.

The *Wilcoxon sign test* is useful in those situations where, for example, we know that a score of 65 on an anxiety scale reflects more anxiety than a lower score of 60 *and* where we also believe that this difference is greater than the difference in anxiety reflected in two other lower scores, say 62 and 60. If this latter determination can be made, we would be throwing away available measurement precision if we were to look at only the direction (more, less, or the same) of difference between pairs of matched cases that constitute two subsamples.

In using the Wilcoxon sign test to evaluate the relative effectiveness of two forms of social work interventions, an assumption is made. If the form of intervention used makes no difference, there will be essentially no difference in regard to the dependent variable among the cases in one group and their counterparts in the other group (the null hypothesis). The ideal situation for using the Wilcoxon sign test would involve the use of perfectly matched pairs of cases (identical twins?). For practical purposes, however, cases that are paired based on a pretest measurement of the dependent variable and/or one or more of the most likely intervening variables is usually sufficient.

Once pairs are identified, one member is randomly assigned to one group; the second goes to the other group. After treatment, the test examines the amount of difference between each pair as well as the direction of the difference. In the process, the differences between pairs are themselves rank ordered. If the preponderance of differences suggests higher scores for one group and the greatest differences are also among those cases, the Wilcoxon sign test is likely to suggest a statistically significant difference between the groups. The null hypothesis can be rejected. The stronger the pattern in this direction, the more likely the null hypothesis can be rejected, and vice versa.

The Wilcoxon sign test can also be used exactly like the correlated groups *t* test when we want to measure the same sample at two different times. However, the Wilcoxon sign test can be used with ordinal level data, whereas the correlated groups *t* test can only be used with interval or ratio level data.

Example.   Linnette is a social work counselor in a student health center. From years of observation, she wondered whether college students having social adjustment problems benefitted more from counseling by untrained student volunteers or by professional social work staff. Using standard intake screening measurements over a one-month period, she identified a group of prospective clients who were all diagnosed as having "moderate social adjustment problems." Before assignment for counseling, she identified 15 matched pairs (matched on such key variables as gender, grade point average, and the like) and randomly assigned one member of each pair to be seen by a student volunteer and the other member to be seen by a social worker. After six one-hour counseling sessions, all clients were administered a scale that measured social adjustment.

The instrument, which Linnette considered an indicator of college students' social adjustment, was deemed to be capable of generating the "ordinal-plus" data required for use of the Wilcoxon sign test. Data were compared for each pair, and the direction and amount of the differences were noted. The differences were rank ordered.

Linnette was pleased to learn that the members of the pairs (students) seen by the social workers scored much better on the scale than did those seen by student volunteers. The Wilcoxon sign test allowed her to feel comfortable in rejecting the null hypothesis. The direction of the differences let her conclude that the clients seen by social workers scored higher on the scale. She was, of course, not ready to discount student volunteers as effective counselors based on this single bit of data, but she wondered whether student volunteers might be used more effectively in working with students having other problems. She also decided to replicate her ministudy, using another measurement of social adjustment to see if consistent findings would be obtained.

PARAMETRIC AND NONPARAMETRIC ANALOGUES

As should be evident by now, there are many statistical tests that can be used to analyze social work research data. They basically fall into two categories—parametric and nonparametric. Which specific test to use is determined by several factors (normally distributed variable within the population from which the sample was drawn, sample size, level of measurement, and so on.). A quick look back at Figure 6.1 reveals the relationship of parametric tests and nonparametric tests. That is, it can be easily seen that the independent groups t test is similar to the Mann-Whitney U test; the correlated groups t test is similar to the Wilcoxon sign test; Pearson's r is similar to Spearman's rho; and so on. It should be noted in Figure 6.1 that there is no parametric counterpart to the chi-square test, Fisher's exact test, or McNemar's test.

All data can be analyzed given the limitations of a given data set. However, the researcher should always determine in advance what statistic will be used to analyze the data when the independent (predictor) variable and the dependent variable are conceptualized and before data collection gets underway. In addition, the researcher should always strive to use the highest level of measurement (operationalization) that is available, given the context of the research situation.

CONCLUDING THOUGHTS

This chapter has presented only a few of the more commonly used nonparametric statistical tests that, along with chi-square, are seen fairly frequently in the social work literature. While not as powerful as their parametric "cousins," they can be made more powerful by increasing sample size. Up to this point, we have looked primarily at univariate (descriptive statistics) and bivariate statistical tests that are designed for examining the simplest of relationships—the relationship between two variables.

As suggested earlier, a common misuse of statistics occurs when a series of univariate analyses is applied when multivariate testing is maybe more appropriate. While an in-depth examination of multivariate statistics is beyond the scope of this introductory book, we think that a conceptual introduction to a few of the more commonly used multivariate methods is appropriate for any student studying statistics. It will be the focus of the next and final chapter.

STUDY QUESTIONS

1. Why would the McNemar's test be especially well suited for social workers who wish to evaluate the impact of a group experience on a stereotype about minorities? Specifically, how could it be used?

2. In what situations can the Fisher's exact test be used when chi-square should not?

3. What conditions for use of the Mann-Whitney U test make it particularly well suited for the individual practitioner who wishes to evaluate the effectiveness of a new treatment method? Provide an example in your discussion.

4. When is the median test preferable to the chi-square test if interval or ratio level data do not meet the necessary criteria for the use of the independent groups t test? Provide an original example of how it could be used.

5. Why is the Kolmogorov-Smirnov test a more comprehensive comparison of ordinal or skewed interval level data drawn from two samples than the median test?

6. What specialized type of sampling is best for the Wilcoxon sign test?

7. What are some of the situations that suggest the need for tests like those described in this chapter rather than those described in Chapters 9 and 10. Provide original examples in your discussion.

8. What do all the tests in this chapter have in common? Discuss in detail.

9. In what ways do all the statistical tests in this chapter and those in Chapters 7, 8, 9, and 10 work in a similar manner?

10. Discuss the possible disadvantages of using statistical tests that are less well known.

11. What do you feel were some of the other variables that Meredith should have taken into account when she interpreted her findings in the hospice example in the discussion of the Spearman's rho test?

12. A number of alternative explanations could have accounted for Meredith's study results. What do you feel these other explanations could have been? Categorize them by (a) rival hypotheses, (b) design bias, and (c) sampling bias.

13. Using other statistics books, find additional nonparametric tests that this chapter did not discuss and add them to Figure 6.1 under the appropriate headings. How are the other tests similar to the ones discussed in this chapter? How are they different?

chapter **12**

Multivariate Options

$\mathbf{C}$hapter 10 alluded to the dangers of using a series of bivariate statistical analyses to test the relationship between a number of independent variables and a single dependent variable. This procedure can result in a Type I error by causing us to stumble onto a spurious "relationship" only because so many possible combinations of variables have been examined. That is one reason to include an overview of multivariate analysis in our discussion of statistical procedures that social workers need to conceptually understand.

However, there also are certain situations where a multivariate analysis is the preferred choice for a data analysis. It is consistent with the multiple causation and systems theories of human behavior that dominate much of the social work literature. Like these theories, multivariate analyses allow us to attempt to sort out the complex interaction of variables that exist within our practice and research milieus.

Multivariate statistical analyses differ from bivariate analyses in two important ways. First, they examine the relationship among three or more variables (that is why it is called *multi*variate). Second, and more importantly, they investigate the relationship among the variables *simultaneously*. What is meant by simultaneously? Suppose we wished to examine the relationship among the predictor variables *per capita income* and *rates of alcoholism* and the criterion variable *incidence of child abuse*. Bivariate analyses might first look at the relationship between income and abuse and then the relationship between alcoholism and abuse. A multivariate analysis, using a single procedure, would look at the total picture, including how income and alcoholism might be related and how all three variables (the two predictor variables and the criterion variable) interact with on another.

Multivariate statistics is the subject of many advanced statistics courses, and an in-depth examination of it is well beyond the scope of this introductory text.

We have included this final chapter because more and more social workers are turning to multivariate analyses to test hypotheses and to describe relationships among variables. While the brief description that follows will not enable readers to perform multivariate statistical analyses, we hope that this chapter (like the previous one) will provide both an appreciation of the richness of this important area of statistics and an impetus to gain more knowledge about it. We will present an overview of a few of the more commonly used multivariate statistical techniques and briefly describe how they can be of use to the social work researcher and practitioner.

TESTS USED FOR CORRELATION AND PREDICTION

Chapters 8 and 9 focused on two related types of statistical analyses—correlation and regression. We will now return to them and discuss ways in which the correlation among more than two variables can be examined using multivariate statistical techniques.

Multiple Correlation

It is useful to know the strength and direction of a relationship between a predictor variable and a criterion variable (bivariate analysis). However, what if two or more predictor variables reflect a high correlation with the same criterion variable? Would we not have an even better chance of explaining the variation within the criterion variable if we could use those two (or more) predictor variables that correlate highly with it? Yes. Logic would tell us that our ability to explain the variation of the criterion variable would be improved if we could determine the *combined* explanation capacity of a group of predictor variables. Multiple correlation (multiple *R*), which we briefly described in Chapter 8, is designed for this task. It determines the degree to which a group of predictor variables does the best job of explaining the variation of a criterion variable. Finding the best combination of predictor variables to do this is not a simple task. We will use an example to see why.

Suppose that we wished to find out how well several predictor variables might explain a person's score on a standardized statistics examination (the criterion variable). We might use four interval level predictor variables: (a) number of statistics courses taken, (b) age, (c) undergraduate GPA; and (d) score on the mathematics section of the SAT.

Let us say that the bivariate correlation between these variables and the criterion variable are, respectively, .45, .26, .51, and .74. To find out which two of these variables might be the best pair to explain a person's score on the statistics examination and how well they would, in combination, explain its variation, we could not simply add the various pairs of bivariate correlations together. Why not? For one thing, adding the correlation coefficients for some of the pairs (e.g., .51 and .74) would produce a correlation coefficient greater than 1.00—and that is impossible.

But more importantly, any two of the predictor variables (or three, or all four of them) will share some of the *same* covariance with the criterion variable. In other words, part of their bivariate correlation with the criterion variable will be "nothing new"—any one predictor variable would account for some of the same variation of the criterion variable as the others. That is because the four predictor variables, in addition to having some correlation with the criterion variable, also have some degree of correlation with each other.

In our example, we would expect there to be a fairly high correlation between the predictor variables of number of statistics courses taken and age, number of statistics courses taken and score on the SAT mathematics section, and so on. Thus, they would not explain a *unique* amount of the variation of the criterion available. In fact, two predictor variables that are highly correlated with each other are unlikely to add much to our ability to explain variation within the criterion variable beyond what either one alone could have done. If they are perfectly correlated (+1.00 or –1.00) they are probably just two measurements of the same variable, and the use of both would be redundant.

Multiple R sorts out the degree to which any one predictor variable accounts for the variation in the criterion variable. It produces what is referred to as a *beta weight* for each predictor variable. Beta weights are derived mathematically based on (a) the correlation between each predictor variable and the criterion variable and (b) the correlations among the various predictor variables. By using both of these, multiple R identifies what is unique within the correlation between a given predictor variable and a criterion variable—that is, the variation in the criterion variable associated with that predictor variable that is not already accounted for by some other predictor variable.

Even though a given predictor variable might have the highest bivariate correlation with the criterion variable (as in the case of the SAT mathematics score in our example), this is no guarantee that its beta weight will also be higher than that of the other predictor variables. If it is highly correlated with the other predictor variables, its beta weight may be relatively low.

Our example used only four predictor variables. But we probably could identify many more interval or ratio level variables that might be expected to correlate fairly highly with scores on a standardized statistics examination. Multiple R can be used with a large number of predictor variables. However, unless we have a considerably larger number of cases than we have variables, we can be led into making too much of the correlation coefficient thus produced. It might be quite high just on the basis of chance. To avoid this problem, multiple R should be used only when the number of cases is substantially larger than the number of predictor variables to be included in the statistical analysis.

There are variations of multiple R, referred to as stepwise procedures or *stepwise regression*, that can be very useful to social work researchers. One of these, referred to as the *step-up procedure*, involves a kind of rank ordering of the predictor variables. First, the predictor variable that is most highly correlated with the criterion variable is identified. Then the next predictor variable that accounts for the most additional unexplained variation of the criterion variable is added to see how much the multiple R correlation coefficient is affected. The process can continue with other predictor variables added (in order of their contribution to the unexplained variance of the criterion variable) until we reach a "point of diminishing returns," that is, until we get down to those predictor variables that are of little value in accounting for any meaningful or useful variance within the criterion variable. The process allows us to narrow our list to only those predictor variables that are of the most important in explaining the variation of the criterion variable.

Another procedure, referred to as the *step-down procedure,* works like the stepwise regression procedure, but in the opposite direction. It begins with all the predictor variables and eliminates them one by one, starting with the one that explains the smallest amount of variation of the criterion variable. Variables previously omitted can be readded along the way to see how the multiple correlation coefficient is affected. Like the step-up procedure previously described, this procedure allows us to "winnow down" the list of predictor variables to the relatively small number that are most useful in predicting the variation of the criterion variable.

As has been mentioned throughout this book, there is a form of statistical analysis for just about any situation. There is yet another situation when still another type of correlation analysis is appropriate. While multiple R is used to obtain the correlation between one criterion variable and a group of predictor variables (sometimes called a *derived variable*), another procedure called *canonical correlation* takes correlation even further. It is used to examine simultaneously the correlation between a weighted *group* of predictor variables (a derived predictor variable) and a weighted *group* of criterion variables (a derived criterion variable). The procedure is used in situations where the criterion concept that we are measuring is somewhat abstract and cannot be adequately measured by using scores for any single variable.

For example, if we wanted to correlate a fairly abstract criterion concept, such as interpersonal skill on the job, we might have to ask co-workers to evaluate persons on several different criterion variables that are indicators of the concept, such as *cooperativeness, communication ability, helpfulness,* and so on. The weighted group of scores on these variables could then be correlated with another weighted group of scores on such predictor variables as *education level, years of work experience, number of siblings,* and so on, using canonical correlation to produce a multiple correlation coefficient. We would then be able to evaluate the multiple correlation coefficient thus produced (or any other one produced by the procedures that we have described above) as to its strength and direction, using the same criteria that are used in evaluating the bivariate correlation coefficients produced by Pearson's *r.* All correlation coefficients, no matter what statistical procedure produced them, are interpreted in the same way, as presented in Chapter 8.

Multiple Regression

It should not be a surprise that both Pearson's *r* (Chapter 8) and simple regression (Chapter 9) both have their multivariate counterparts. Nor should it be surprising that multiple correlation and multiple regression have many similarities. However, they have both theoretical and actual differences.

In theory, all correlation analyses (including multiple R) are designed to analyze data where the researchers have no control over the distribution of the variables (random variables) within the population. Researchers simply measure the predictor and criterion variables ex post facto, after they have already been distributed among cases or objects. Consequently, a correlation coefficient is just an indication of how these variables covary naturally, that is, without any interference by the researchers. No matter how high the correlation coefficient produced, it would be presumptuous to imply that the predictor variable in any way caused the variations in the values of the criterion variable.

Regression analysis, however, was designed for situations where we have experimental data—where the predictor variables are either introduced or directly manipulated by the researchers. That is the way that correlation and regression are supposed to be used. In reality, regression analysis is now also used in many situations where we have only random variables, whose distribution is beyond the control of the researchers. This causes no major problems, so long as we are confident that the measurements of the predictor variables are accurate. The message to the social work researcher is this: In those relatively rare situations in which experimental designs are employed and interval or ratio level measurement exists for both predictor and criterion variables, regression analysis (simple or multiple) is the procedure of choice. If the predictor variables are really random variables, either correlation or regression can be used, depending on the information sought.

Multiple regression is used to help us predict the value of an interval or ratio level criterion variable using two or more interval or ratio predictor variables. Multiple R produces a correlation coefficient that gives us a good idea of the degree of correlation between the predictor variables and the criterion variable. When squared (R^2) it can tell us approximately what percentage of the variation within values of the criterion variable can be explained by the predictor variables, viewed as a group. Multiple regression does more. It can, through the use of a regression equation, help us to estimate the actual value of a criterion variable, knowing the corresponding values for a group of predictor variables.

Like simple linear regression, multiple regression is used to: (a) decide whether an apparent relationship between the predictor variables and the criterion variable is likely to be real or the work of chance, (b) present any such relationship as a mathematical equation, and (c) indicate how accurate the equation is as a predictor of values of the criterion variable. But what makes multiple regression different from simple linear regression is that it also provides an assessment of the relative value (i.e., respective contributions) of different predictor variables for predicting values of the criterion variable.

The formula for multiple regression is similar to that of simple linear regression, but not surprisingly it is more complex. It certainly is a job for a computer. Like the formula for multiple R, it takes into consideration the fact that the predictor variables are not only correlated with the criterion variable, they are also correlated to a greater or lesser degree with each other. We could not just simply add up their individual prediction powers, because some of the prediction power of one predictor variable duplicates the power within another predictor variable.

The regression counterpart of partial r, beta weights, are referred to as *beta coefficients* or, sometimes, as *partial regression coefficients.* Like beta weights, they reflect the correlation between a given predictor variable and other predictor variables as well as between it and the criterion variable. There are other similarities between multiple R and multiple regression as well. Stepwise procedures are available to help us select the smallest number of predictor variables while not sacrificing too much in our ability to predict. Like multiple R, this can be accomplished by using either a step-down or step-up method, that is, by systematically adding or dropping different predictor variables. Statistical software packages can quickly provide us with the best (in terms of prediction of the criterion variable) combination of two, three, four, or more predictor variables that will produce the best multiple regression equation.

With a little ingenuity, multiple regression analyses even can be used to include as predictor variables some variables that are inherently nominal or ordinal. We can do this by creating dummy variables (see Chapter 1). For example, we could create a dummy variable, *current marriedness,* with a value of 0 for those people who are not married and a value of 1 for those who are. The variable *marital status,* which is generally nominal, would then become ratio, with the 0 representing a state of no measurable quantity of the variable. Of course, in so doing, we would have lumped all cases, such as those divorced, widowed, unmarried with a permanent different-sex or same-sex partner, and so on, into one value category. We will have lost some measurement precision, but the opportunity to perform relatively high powered multivariate analyses using marital status along with several interval or ratio variables may be worth the loss.

Like any method of statistical analysis, multiple regression can produce results that, if we are not careful, can lead us into Type I or Type II errors. An advanced statistics text or a consultant can help us to avoid such pitfalls as collinearity (two highly correlated predictor variables that can distort the results of an analysis), self-fulfilling prophecies, or regression to the mean, which can occur unless proper precautions are taken.

Discriminant Analysis

In social work research we often find ourselves with more than one predictor variable and a criterion variable that is only at the nominal or ordinal level of measurement. We want to increase our ability to predict the values of such criterion variables such as *recidivism* (abused again/did not abuse again), *rehospitalization* (rehospitalized/not rehospitalized), or *electoral involvement* (voted/did not vote) by identifying a group of interval or ratio level predictor variables that will do the best job in predicting the variation of the criterion available. The criterion variable may be dichotomous, as in the previous examples, or it may contain three or more such categories (values), as in the case of a variable such as *religious affiliation* (Christian/Muslim/Jewish/Buddhist/Hindu, and so on). We may want to know what "kind" of people select the various denominations, that is, what group of predictor variables might be most helpful in predicting one's religious affiliation. In either situation, discriminant analysis may be the procedure that is indicated.

Like multiple regression, multivariate discriminant analysis creates a derived variable (in this case called the *discriminant function*) from the weighted values of several predictor variables. Stepwise variations of the procedure likewise are available. Statistical software programs also can produce an interesting and useful table (often called a *confusion matrix*). It displays the type and number of errors that a given group of predictor variables produce. It includes the number of cases that actually were in a given category of the criterion value and the number that the predictor variables would have predicted would be there. It thus allows the researcher to know the likely number of prediction errors in any direction if a certain group of predictor variables were to be used, and to assess whether that number would be acceptably low.

Discriminant analysis is closely related to multiple regression. It is less powerful than multiple regression, in a statistical sense, because it requires only nominal or ordinal level measurement of the criterion variable. While it *can* be used with an interval or ratio level criterion variable, it should not. Why throw away measurement precision and statistical power unnecessarily? To use discriminant analysis rather than multiple regression with an interval or ratio level predictor variable would be the equivalent of using chi-square when the requirements for an independent groups *t* test can be met.

OTHER USEFUL PROCEDURES

We will briefly mention three more forms of multivariate analyses that are gaining popularity among social work researchers. The first two have in common the fact that they tend to be used more for tasks of research design (instrument construction and conceptualization of variables, respectively) than for testing the relationships among variables. The third procedure returns our focus to what has been a primary concern of the latter part of this book, hypothesis testing.

Factor Analysis

Factor analysis is a mathematically sophisticated form of data reduction. It provides a way of reducing many items contained in a measuring instrument to a smaller number that are believed to measure essentially the same variable.

In designing multi-item measuring instruments to measure a variable, factor analysis can be used to reduce the number of items in the instrument without losing the instrument's measurement capacity. When we first design a measuring instrument, we may have many more items (or questions) than we ultimately hope to have. We may have included virtually any item that we think might measure the variable that we are trying to measure. We would like to know which items are redundant, that is, two or more items that measure the same thing, so that we can eliminate some of them. Factor analysis can help us with this task.

Factor analysis relies heavily on the concept of correlation. It examines to what degree peoples' responses to one item correlate with their responses to other items. Two or more items that produce the same or nearly the same responses from large numbers of research participants are assumed to be measuring the same variable. Those items that produce dissimilar responses are assumed to be measuring something else. For example, in an instrument measuring attentiveness in class, one item might ask how frequently a student has fallen asleep in class. Another item might ask how often the student yawns in class.

If, in a large sample of students, there is a strong pattern (positive correlation) of responses wherein individual students tend to respond either "frequently" to both items or "never" to both items, it may be assumed that the two items measure the same dimension of attentiveness. If, however, individual students tended to give very different responses to the two items (which would be surprising!), we might conclude that the two items measure two different dimensions of attentiveness.

Using a variation of correlation analysis, factor analysis groups items (which can be thought of loosely as variables) based on the responses to them within a large data set. Returning to our example, it might list four different items in one group because responses to them were similar among individuals. This would indicate that all four items probably are measurements of the same *factor* (the same abstract underlying dimension of the variable that we are attempting to measure). We could then look at the four items themselves to determine what they have in common.

If two of the four items ask about head nodding and resting one's head on one's hand, we may decide to label this factor something like "physical manifestations of attentiveness." We may decide that any one of the items may be sufficient and the other three can be deleted, or we may wish to leave two, three, or all four of them in the next version of the scale as a now-identified subscale.

The factors derived from factor analysis are dimensions that are *independent* of other factors. This does not mean, however, that an item will show up in only one group of items (a factor) formed by a factor analysis. An item may be a part of two (or even more) factors. "Judgment calls" must be made when factor analyses are used. Questions of how many items or how many factors to retain are complicated ones.

Among other considerations, we may ask how long a scale can be before respondents refuse to complete it. One product of a factor analysis is helpful in this regard. It is an *eigenvalue*, which is a number that corresponds to the number of items that a given factor represents. For example, if we had a factor with an eigenvalue of 4.3, the factor would be the equivalent of 4.3 items. A factor with an eigenvalue of only .8 would be the equivalent of less than one item in terms of its contribution to measurement of a variable. In the interest of efficiency, a scale usually would retain only those factors that have relatively high eigenvalues. A value such as .8 would suggest that retaining its corresponding factor would only "clutter" the scale and make it unnecessarily long and cumbersome. It would add little to the measurement of the variable.

There are a number of other features to factor analyses that allow us to construct just about any form of measuring instrument that we require. Factors can be refined using a process called *factor rotation*. While the process is complicated and has many different variations, all of them are designed to help us to develop more precise meanings of the factors within any given data set.

A second practical use of factor analysis, in addition to instrument building, is for the preliminary screening of large groups of variables prior to multiple regression or discriminant analysis. Redundant predictor variables, which appear as part of the same factors, can be identified, and we may be able to eliminate all but one, thus making the job of either type of analysis much easier.

Factor analyses have many practical uses in fields such as business and marketing, but they are useful in social work as well. Restaurant owners sometimes do a factor analysis of their customers' food preferences to try to identify menu items that are redundant (i.e., may appeal to the same customers). Once identified, some of these items can be eliminated and replaced with others that have the potential to broaden the restaurant's potential clientele. Marketing managers sometimes perform factor analyses to determine what factors lead people to buy a certain product. Then they develop magazine advertisements based upon the results of the analyses. The advertisements are written carefully to avoid redundancy (the same factors) while covering all identified factors.

A social work manager might use a factor analysis for making similar everyday, practical decisions. For example, discharge summary records data could be used to determine which factors constitute client satisfaction. Then staff development sessions could be used to build in all factors while not engaging in unnecessary and inefficient redundancy ("overkill").

Factor analyses can be very helpful in improving many forms of evaluation in social work. For example, they can be used to refine instruments used by students to evaluate professors, eliminating redundancy that places disproportionate weight on a single dimension (factor) of teaching effectiveness such as "niceness" or "preparedness."

Cluster Analysis

While factor analyses perform the task of clustering variables or items into factors, *cluster analysis,* also referred to as *taxonomy analysis,* is used primarily to cluster cases. It is used to form subsamples of cases based on the fact that they have similar measurements (values) in relation to certain variables. Consequently, cluster analyses usually are used more to make hypothesis testing possible (by creating value categories of variables) rather than for conducting hypothesis testing itself.

While cluster analyses have some similarities to discriminant analyses, there is one important difference. When we conduct a discriminant analysis we begin with one nominal or ordinal level variable that is already "sorted" into categories (values) and attempt to learn how the categories differ. However, in cluster analysis we lack clearly defined groups (categories) when we begin. The goal of a cluster analysis is to form groups or subsamples of cases that are similar to each other but different from cases in other subsamples in relation to certain variables.

To illustrate this difference, a little earlier we described how we might conduct a discriminant analysis using people's religious affiliations as our nominal level variable and proceed to try to identify those variables (e.g., income, education level, age) that might distinguish one religious affiliation from another. However, we might use a cluster analysis to attempt to create groups based upon similar beliefs about a divine being or other religious beliefs. After the cluster analysis, each group formed would probably contain persons who represent a variety of religious affiliations but who hold similar beliefs. The beliefs of individuals in any one group would be different from those of persons in the other groups. Yet a member of a group may have the same religious affiliation as a member of another group.

The example above suggests an important use of a cluster analysis for the social worker. Sometimes the similarity of people's attitudes, beliefs, values, or behaviors is of more utility to us than their label or diagnostic category. For example, it might be useful to employ cluster analysis to group clients with psychological disabilities based upon their frequency of several self-defeating behaviors, rather than upon their DSM-4 diagnosis. We could then use the clusters (groups) thus formed to constitute different treatment groups that would have quite different therapeutic goals.

Multiple Analysis of Variance

In Chapter 10 we discussed one-way analysis of variance (*ANOVA*) and briefly mentioned various factorial designs that are used when we have more than one nominal or ordinal level independent variables. *Multiple analysis of variance* (*MANOVA*) is designed for situations where we wish to test for relationships that involve more than one dependent variable. Like *ANOVA*, *MANOVA* entails a comparison of means, but it compares the means of dependent variables for one category (value) of the nominal or ordinal value with means of dependent variables for the other categories of that variable. Thus, it compares groups (sets) of means compiled from several interval or ratio level variables.

In situations where *MANOVA* is used, a series of *ANOVA* procedures could have been used instead. But to do so would fail to address the importance of intercorrelation among the two or more dependent variables. The formula for *MANOVA* takes this phenomenon into consideration.

CONCLUDING THOUGHTS

We have come a long way in our discussion of statistics. Like any journey attempting to cover a great amount in a short period of time, ours may have included some bad moments for the reader. You may have gotten lost for a while or may have wished to remain longer in one place.

Our role as authors has been that of tour guide. Given space and time constraints, we have pointed out those forms of statistical analyses that we believe to be of the most value and interest to social workers. We have chosen to emphasize conceptual understanding and appreciation for the use of various forms of analyses rather than their mathematical underpinnings.

We provided a more detailed discussion in early chapters since we believe that certain basic ideas are essential for understanding more complicated concepts, or at least to know what questions to ask. In the last two chapters, we have clearly done little more than scratch the surface. It is hoped the reader has emerged from them with a general understanding of the procedures and the situations in which they are used and is now convinced that there is a procedure for any data analysis requirement that we can ever imagine. If you wish to learn more about the field of statistical analysis through advanced study, better yet! We have chosen to sacrifice comprehensiveness for simplicity to accomplish our goal of providing a "reader friendly" introduction to statistics.

STUDY QUESTIONS

1. In what ways do multivariate analyses differ from bivariate analyses? Why are they less likely to cause a researcher to commit a Type I error?

2. Find an article in a professional social work journal that used multiple correlation as the method of data analysis. What were the independent variables? What conclusions did the author generate from the study? Do you feel the study was relevant to your future practice? Why or why not? What other independent variables could the author have used?

3. When using multiple correlation, why can we not simply add together the individual bivariate correlation coefficients between two predictor variables and the criterion variable to find their *combined* correlation with the criterion variable? Explain.

4. Which form of multivariate analysis is designed to help us predict the value of an interval or ratio level criterion variable from the measurements of two or more interval or ratio level predictor variables that have been introduced or manipulated by the researcher?

5. Describe how dummy variables can be used to perform multiple regression using a nominal level dependent variable. Provide an example.

6. How do step-up and step-down regression differ from each other? Describe how each works.

7. Which multivariate statistic could be used to create treatment groups of clients that are homogeneous in relation to such variables as *attitudes toward abortion, attitudes toward corporal punishment,* and so on?

8. Which multivariate statistic examines the relationship between more than one predictor variable and a nominal or ordinal level criterion variable? Provide an example of its use.

9. Explain how factor analyses use the concept of reliability to reduce a large number of variables to a small number of factors. What is an eigenvalue, and what does it represent in a factor analysis?

10. Which forms of multivariate analyses described in this chapter are used primarily for prediction or hypothesis testing? Which ones are frequently used for research design tasks?

Appendixes

APPENDIX A
Areas of the Normal Curve

Area Under the Normal Curve Between Mean and z Score

z	.00	.01	.02	.03	.04	.05	.06	.07	.08	.09
0.0	00.00	00.40	00.80	01.20	01.60	01.99	02.39	02.79	03.19	03.59
0.1	03.98	04.38	04.78	05.17	05.57	05.96	06.36	06.75	07.14	07.53
0.2	07.93	08.32	08.71	09.10	09.48	09.87	10.26	10.64	11.03	11.41
0.3	11.79	12.17	12.55	12.93	13.31	13.68	14.06	14.43	14.80	15.17
0.4	15.54	15.91	16.28	16.64	17.00	17.36	17.72	18.08	18.44	18.79
0.5	19.15	19.50	19.85	20.19	20.54	20.88	21.23	21.57	21.90	22.24
0.6	22.57	22.91	23.24	23.57	23.89	24.22	24.54	24.86	25.17	25.49
0.7	25.80	26.11	26.42	26.73	27.04	27.34	27.64	27.94	28.23	28.52
0.8	28.81	29.10	29.39	29.67	29.95	30.23	30.51	30.78	31.06	31.33
0.9	31.59	31.86	32.12	32.38	32.64	32.90	33.15	33.40	33.65	33.89
1.0	34.13	34.38	34.61	34.85	35.08	35.31	35.54	35.77	35.99	36.21
1.1	36.43	36.65	36.86	37.08	37.29	37.49	37.70	37.90	38.10	38.30
1.2	38.49	38.69	38.88	39.07	39.25	39.44	39.62	39.80	39.97	40.15
1.3	40.32	40.49	40.66	40.82	40.99	41.15	41.31	41.47	41.62	41.77
1.4	41.92	42.07	42.22	42.36	42.51	42.65	42.79	42.92	43.06	43.19
1.5	43.32	43.45	43.57	43.70	43.83	43.94	44.06	44.18	44.29	44.41
1.6	44.52	44.63	44.74	44.84	44.95	45.05	45.15	45.25	45.35	45.45
1.7	45.54	45.64	45.73	45.82	45.91	45.99	46.08	46.16	46.25	46.33
1.8	46.41	46.49	46.56	46.64	46.71	46.78	46.86	46.93	46.99	47.06
1.9	47.13	47.19	47.26	47.32	47.38	47.44	47.50	47.56	47.61	47.67
2.0	47.72	47.78	47.83	47.88	47.93	47.98	48.03	48.08	48.12	48.17
2.1	48.21	48.26	48.30	48.34	48.38	48.42	48.46	48.50	48.54	48.57
2.2	48.61	48.64	48.68	48.71	48.75	48.78	48.81	48.84	48.87	48.90
2.3	48.93	48.96	48.98	49.01	49.04	49.06	49.09	49.11	49.13	49.16
2.4	49.18	49.20	49.22	49.25	49.27	49.29	49.31	49.32	49.34	49.36
2.5	49.38	49.40	49.41	49.43	49.45	49.46	49.48	49.49	49.51	49.52
2.6	49.53	49.55	49.56	49.57	49.59	49.60	49.61	49.62	49.63	49.64
2.7	49.65	49.66	49.67	49.68	49.69	49.70	49.71	49.72	49.73	49.74
2.8	49.74	49.75	49.76	49.77	49.77	49.78	49.79	49.79	49.80	49.81
2.9	49.81	49.82	49.82	49.83	49.84	49.84	49.85	49.85	49.86	49.86
3.0	49.87									
3.5	49.98									
4.0	49.997									
5.0	49.99997									

Source: The original data for Appendix A came from *Tables for Statisticians and Biometricians*, edited by K. Pearson, published by the Imperial College of Science and Technology, and are used here by permission of the Biometrika trustees. The adaptation of these data is taken from E. L. Lindquist, *A First Course in Statistics* (revised edition), with permission of the publisher, Houghton Mifflin Company.

APPENDIX B
Critical Values of χ^2

	Level of Significance for a One-Tailed Test					
	.10	.05	.025	.01	.005	.0005
	Level of Significance for a Two-Tailed Test					
df	.20	.10	.05	.02	.01	.001
1	1.64	2.71	3.84	5.41	6.64	10.83
2	3.22	4.60	5.99	7.82	9.21	13.82
3	4.64	6.25	7.82	9.84	11.34	16.27
4	5.99	7.78	9.49	11.67	13.28	18.46
5	7.29	9.24	11.07	13.39	15.09	20.52
6	8.56	10.64	12.59	15.03	16.81	22.46
7	9.80	12.02	14.07	16.62	18.48	24.32
8	11.03	13.36	15.51	18.17	20.09	26.12
9	12.24	14.68	16.92	19.68	21.67	27.88
10	13.44	15.99	18.31	21.16	23.21	29.59
11	14.63	17.28	19.68	22.62	24.72	31.26
12	15.81	18.55	21.03	24.05	26.22	32.91
13	16.98	19.81	22.36	25.47	27.69	34.53
14	18.15	21.06	23.68	26.87	29.14	36.12
15	19.31	22.31	25.00	28.26	30.58	37.70
16	20.46	23.54	26.30	29.63	32.00	39.29
17	21.62	24.77	27.59	31.00	33.41	40.75
18	22.76	25.99	28.87	32.35	34.80	42.31
19	23.90	27.20	30.14	33.69	36.19	43.82
20	25.04	28.41	31.41	35.02	37.57	45.32
21	26.17	29.62	32.67	36.34	38.93	46.80
22	27.30	30.81	33.92	37.66	40.29	48.27
23	28.43	32.01	35.17	38.97	41.64	49.73
24	29.55	33.20	36.42	40.27	42.98	51.18
25	30.68	34.38	37.65	41.57	44.31	52.62
26	31.80	35.56	38.88	42.86	45.64	54.05
27	32.91	36.74	40.11	44.14	46.94	55.48
28	34.03	37.92	41.34	45.42	48.28	56.89
29	35.14	39.09	42.69	46.69	49.59	58.30
30	36.25	40.26	43.77	47.96	50.89	59.70
32	38.47	42.59	46.19	50.49	53.49	62.49
34	40.68	44.90	48.60	53.00	56.06	65.25
36	42.88	47.21	51.00	55.49	58.62	67.99
38	45.08	49.51	53.38	57.97	61.16	70.70
40	47.27	51.81	55.76	60.44	63.69	73.40
44	51.64	56.37	60.48	65.34	68.71	78.75
48	55.99	60.91	65.17	70.20	73.68	84.04
52	60.33	65.42	69.83	75.02	78.62	89.27
56	64.66	69.92	74.47	79.82	83.51	94.46
60	68.97	74.40	79.08	84.58	88.38	99.61

Source: From Table IV of R. A. Fisher and F. Yates, *Statistical Tables for Biological, Agricultural and Medical Research*, published by Longman Group, Ltd., London (previously published by Oliver and Boyd, Ltd., Edinburgh) and by permission of the authors and publishers.

APPENDIX C
Critical Values of r

	Level of Significance for a One-Tailed Test				
	.05	.025	.01	.005	.0005
	Level of Significance for a Two-Tailed Test				
N	.10	.05	.02	.01	.001
5	.8054	.8783	.9343	.9587	.9912
6	.7293	.8114	.8822	.9172	.9741
7	.6694	.7545	.8329	.8745	.9507
8	.6215	.7067	.7887	.8343	.9249
9	.5822	.6664	.7498	.7977	.8982
10	.5494	.6319	.7155	.7646	.8721
11	.5214	.6021	.6851	.7348	.8471
12	.4973	.5760	.6581	.7079	.8233
13	.4762	.5529	.6339	.6835	.8010
14	.4575	.5324	.6120	.6614	.7800
15	.4409	.5139	.5923	.6411	.7603
16	.4259	.4973	.5742	.6226	.7420
17	.4124	.4821	.5577	.6055	.7246
18	.4000	.4683	.5425	.5897	.7084
19	.3887	.4555	.5285	.5751	.6932
20	.3783	.4438	.5155	.5614	.6787
21	.3687	.4329	.5034	.5487	.6652
22	.3598	.4227	.4921	.5368	.6524
27	.3233	.3809	.4451	.4869	.5974
32	.2960	.3494	.4093	.4487	.5541
37	.2746	.3246	.3810	.4182	.5189
42	.2573	.3044	.3578	.3932	.4896
47	.2428	.2875	.3384	.3721	.4648
52	.2306	.2732	.3218	.3541	.4433
62	.2108	.2500	.2948	.3248	.4078
72	.1954	.2319	.2737	.3017	.3799
82	.1829	.2172	.2565	.2830	.3568
92	.1726	.2050	.2422	.2673	.3375
102	.1638	.1946	.2301	.2540	.3211

Source: From Table VII of R. A. Fisher and F. Yates, *Statistical Tables for Biological, Agricultural and Medical Research*, published by Longman Group, Ltd., London (previously published by Oliver and Boyd, Ltd., Edinburgh) and by permission of the authors and publishers.

APPENDIX D
Critical Values of *t*

	Level of Significance for a One-Tailed Test					
	.10	.05	.025	.01	.005	.0005
	Level of Significance for a Two-Tailed Test					
df	.20	.10	.05	.02	.01	.001
1	3.078	6.314	12.706	31.821	63.657	636.619
2	1.886	2.920	4.303	6.965	9.925	31.598
3	1.638	2.353	3.182	4.541	5.841	12.941
4	1.533	2.132	2.776	3.747	4.604	8.610
5	1.476	2.015	2.571	3.365	4.032	6.859
6	1.440	1.943	2.447	3.143	3.707	5.959
7	1.415	1.895	2.365	2.998	3.499	5.405
8	1.397	1.860	2.306	2.896	3.355	5.041
9	1.383	1.833	2.262	2.821	3.250	4.781
10	1.372	1.812	2.228	2.764	3.169	4.587
11	1.363	1.796	2.201	2.718	3.106	4.437
12	1.356	1.782	2.179	2.681	3.055	4.318
13	1.350	1.771	2.160	2.650	3.012	4.221
14	1.345	1.761	2.145	2.624	2.977	4.140
15	1.341	1.753	2.131	2.602	2.947	4.073
16	1.337	1.746	2.120	2.583	2.921	4.015
17	1.333	1.740	2.110	2.567	2.898	3.965
18	1.330	1.734	2.101	2.552	2.878	3.922
19	1.328	1.729	2.093	2.539	2.861	3.883
20	1.325	1.725	2.086	2.528	2.845	3.850
21	1.323	1.721	2.080	2.518	2.831	3.819
22	1.321	1.717	2.074	2.508	2.819	3.792
23	1.319	1.714	2.069	2.500	2.807	3.767
24	1.318	1.711	2.064	2.492	2.797	3.745
25	1.316	1.708	2.060	2.485	2.787	3.725
26	1.315	1.706	2.056	2.479	2.779	3.707
27	1.314	1.703	2.052	2.473	2.771	3.690
28	1.313	1.701	2.048	2.467	2.763	3.674
29	1.311	1.699	2.045	2.462	2.756	3.659
30	1.310	1.697	2.042	2.457	2.750	3.646
40	1.303	1.684	2.021	2.423	2.704	3.551
60	1.296	1.671	2.000	2.390	2.660	3.460
120	1.289	1.658	1.980	2.358	2.617	3.373

Source: From Table III of R. A. Fisher and F. Yates, *Statistical Tables for Biological, Agricultural and Medical Research*, published by Longman Group, Ltd., London (previously published by Oliver and Boyd, Ltd., Edinburgh) and by permission of the authors and publishers.

Selected Glossary

This glossary is designed as a general reference for the student of statistics. Many of the terms are discussed in detail in the text. Others, which frequently appear elsewhere in the literature of statistical analysis, also have been included.

Abscissa. See **Horizontal axis.**

Absolute frequency distribution. A table that displays the frequencies for various measurements of a variable.

Acceptance region. The outcome of a statistical test that leads to the acceptance of the null hypothesis.

Allowance factor. Used in constructing confidence intervals, it is the distance (on the measurement scale) between the sample statistic and the limits of the interval. We both add and subtract the allowance factor to find, respectively, the upper and lower limits of the confidence interval.

Alpha error. See **Type I error.**

Alternative hypothesis. See **Research hypothesis.**

Analysis of covariance. A method of statistical control through which scores on the dependent variable are adjusted according to scores on a related variable.

Analysis of variance. A statistical technique by which it is possible to partition the variance in a distribution of scores according to separate sources or factors; a statistical measure to test the differences between the means of two or more groups; sometimes referred to as *ANOVA.*

ANOVA. The abbreviation for the statistical procedure known as **analysis of variance.**

Antecedent variable. A variable that precedes the introduction of the dependent variable.

A priori probability. The probability of a future event calculated from prior knowledge of the number of possible outcomes and their relative frequencies.

Arithmetic mean. See **Mean.**

Axes. Reference lines that delineate the two (or sometimes three) dimensions of a graph; the horizontal and vertical lines in a graph upon which values of a measurement or the corresponding frequencies are plotted.

Bar graph. A graphical technique of descriptive statistics that uses the heights of separated bars to show how often each score occurs; graphical representation of a frequency distribution table in which each measurement category is represented by a bar that extends to the appropriate distance in the frequency dimension; usually has spaces between bars to represent nominal level data.

Beta error. See **Type II error.**

Biased sample. A sample unintentionally selected in such a way that some members of the population are more likely than others to be picked for sample membership; if we wish to make generalizations about the population based on sample observations, it is desirable to avoid biased samples.

Bimodal distribution. A frequency distribution with two modes reflecting equal or nearly equal frequencies.

Binary variable. A dichotomous variable whose values are 0 (reflecting absence of any quantity of the variable) and 1 (reflecting presence of the variable).

Bivariate analysis. A statistical analysis of the relationship between two variables.

Causality. A relationship of cause and effect; the effect will invariably occur when the cause is present; causality is usually statistical; changes in the causal variable (independent variable) will alter values of the affected variable (dependent variable), on the average.

Causal relationship. A relationship between two variables for which we can say that the presence or absence of one variable determines the presence or absence of the other or that values of one variable result in specific values of the other variable.

Cell. A compartment in a matrix or table, such as in a cross-tabulation table.

Central tendency. A typical value for a variable within a data set; one of several descriptive statistics used to reflect a middle value within an array of case values.

Chance. The probability of an event occurring because of some random variation. Sometimes referred to as **sampling bias.**

Chi-square, or goodness-of-fit, test. A technique of inferential statistics used to decide whether a sample with a given frequency distribution could have occurred by chance from a population with a known frequency distribution (or known percentage composition); a nonparametric statistic that allows us to decide whether observed frequencies are essentially equal to or significantly different from expected frequencies.

Chi-square table. See **Cross-tabulation table.**

Class boundary. Dividing point between two cells in a frequency histogram.

Classes. Cells of a frequency histogram.

Class frequency. Number of observations falling in a class (referring to a frequency histogram).

Class interval. The interval between the highest and lowest values in each category of a grouped frequency distribution or histogram.

Cluster analysis. A statistical procedure that, among its other uses, groups together those instruments that measure the same constructs.

Coding. The act of categorizing raw data into groups or giving the data numerical values.

Coding frame. A statement of what is to be coded and how it is to be coded to prepare data for analyses.

Coefficient of determination. The proportion of variation in a scattergram that is explained—that is, the proportion of variation of the criterion variable accounted for by the predictor variables; the coefficient of determination is equal to r^2, where r is the Pearson's r for the two variables.

Coefficient of nondetermination. Equal to $1 - r^2$; the proportion of the variation of the criterion variable that is not accounted for by the predictor variable.

Conceptualization. The first step in the measurement process, in which the researcher selects which variables needed to be measured; delineating the exact meaning of the independent and dependent variables.

Concomitant variation. The case in which two variables vary together; individuals who differ with respect to Variable X will also differ with respect to Variable Y.

Confidence coefficient. Probability that an interval estimate (a confidence interval) will enclose the parameter of interest.

Confidence interval. A range of values within which we are willing to assert with a specified level of confidence that an unknown parameter value lies; computed from sample statistics, the width of the confidence interval depends on the rejection level stated, the sample size, and the variability within the sample.

Confidence limits. Upper and lower boundaries of confidence intervals.

Confounding variables. Variables operating in a specific situation in such a way that their effects cannot be separated; they occur when the effects of an extraneous variable cannot be separated from the effects of the dependent variable; the effects of the extraneous variable thus confound the interpretation of research results.

Confusion matrix. A table that displays the type and number of errors that a given group of predictor variables have produced; a product of discriminant analysis.

Constant. A characteristic that has the same value for all individuals in a research study.

Constant error. A deviation from a true measure resulting from some factor that systematically affects the characteristic being measured or the process of measurement.

Contingency table. See **Cross-tabulation table.**

Continuous random variable. A random variable that may theoretically assume any value between two points on the measurement scale; it can thus have an infinite number of possible values between those points.

Control group. A group of people who do not receive the experimental treatment; a group used for comparison purposes; those people to whom no experimental stimulus is administered but who resemble members of the experimentalgroup in all other respects; in an experimental research design, a group in which the independent variable is left unchanged; serves as a reference to compare the effect of manipulating the independent variable in the experimental group(s).

Control variable. A variable, other than the independent variable(s) of primary interest, whose effects we can determine; an intervening variable that has been controlled for in the research design; a variable that is included in designs as an independent variable for the purpose of explaining (controlling) variation.

Correlated groups *t* test. A hypothesis-testing procedure used to decide whether two given dependent samples could have occurred by chance; sometimes referred to as dependent *t* tests. See also *t* **test.**

Correlated variables. Variables whose values are associated; values of one variable tend to be associated in a systematic way with values in the others.

Correlational analyses. Statistical methods that allow us to discover, describe, and measure the strength and direction of associations between and among variables; include the various techniques of computing correlation coefficients and regression analyses.

Correlation coefficient. A single statistic that indicates both the strength and direction of the relationship between two ordinal, interval, or ratio level variables; correlation coefficients have values between +1 and –1, with positive values indicating positive relationships and negative values indicating negative relationships; two commonly used correlation coefficients are the Pearson's product-moment correlation coefficient (Pearson's *r*) and the Spearman's rho.

Correlation matrix. A table used to display the correlations among three or more pairs of variables.

Covariate. The measure used in an analysis of covariance for adjusting the scores of the dependent variable.

Criterion variable. The variable whose values are predicted from measurements of the predictor variable.

Critical region. A set of outcomes of a statistical test that leads to the rejection of the null hypothesis.

Critical value. A value of a test statistic that demarcates the region of rejection and that is thus used as a criterion for statistical significance in hypothesis testing; the value of the statistic that marks the significance level.

Cross-break table. See **Cross-tabulation table.**

Cross-tabulation table. A table showing the joint frequency distribution of two or more nominal level variables; presents how often each combination of values of each variable occurs; the entries in the table show the number of observations falling into the cells.

Cumulative frequency distribution. A frequency distribution that gives the number of scores that occur at or below each value of a variable.

Cumulative frequency polygon. A frequency polygon that shows how often scores occur at or below each value of a variable.

Cumulative percentage distribution. A table that shows what percentage of scores occur at or below each value of a variable.

Cumulative proportion graph. A graph in which one axis represents values of a variable and the other represents the proportion of the distribution that falls below those values (i.e., their cumulative proportions); when the data have been grouped, each point on the graph is plotted over the upper true limit of the interval it represents; each point thus represents the proportion of the observations falling at or below that interval; under certain conditions, the line connecting points on the graph approximates a curve known as an *ogive*.

Cumulative proportion table. A summary table of a group of observations that has one column listing values of a variable and another column indicating the proportion of the distribution that falls at or below each value; when the data have been grouped, the table lists intervals on the measurement scale rather than individual values.

Curvilinear correlation. A relationship between variables that, if displayed using a scattergram, would form one or more curves; a relationship between two variables that is not linear.

Data. The numbers, or scores, generated by a research study; the word *data* is plural.

Datum. Singular of **data.**

Decision rule. A statement that we use (in testing a hypothesis) to choose between the null or the research hypothesis; the decision rule indicates the range(s) of values of the observed statistic that leads to the rejection of the null hypothesis.

Degrees of freedom. A characteristic of the sample statistic that determines the appropriate sampling distribution; the number of ways in which the data are free to vary; the number of observations minus the number of restrictions placed on the data; a number related to the sample size in a way that depends on the particular statistical technique employed; in many statistical tests, degree of freedom, or *df*, is needed in order to look up critical values.

Dependent events. Events that influence the probability of occurrence of each other.

Dependent variable. The variable that we do not directly introduce or manipulate; after the different levels of the independent variable have been administered, all research participants are measured, in the same way, on the same dependent variable; a variable in which the changes are results of the level or amount of the independent variable(s); also, the variable of most interest to the researcher; when used with correlation or regression it is referred to as the **criterion variable.**

Descriptive statistics. Methods used for summarizing and describing data in a clear and precise manner; strictly speaking, descriptive statistics apply only to the people (or objects) actually observed; methods for data reduction.

Design bias. Any effect that systematically distorts the outcome of a research study so that the results are not representative of the phenomenon under investigation.

Deviation from the mean. The distance of a single score from the mean of the distribution from which the scores come.

Deviation score. The difference between the mean of a distribution and an individual score of that distribution; deviation scores are always found by subtracting the mean from the score; a positive value indicates a score above the mean; a negative value indicates a score below the mean.

Dichotomous variable. A variable that can take on only one of two values.

Directional hypothesis. A hypothesis stated in such a manner that the direction of the relationship between variables is hypothesized for the results; it uses a statistical test with only one region of rejection, that is, a one-tailed test; a directional test is called for only when certain assumptions can be made; because the region of rejection is located entirely at one end of the distribution in a directional test, fewer deviant values of the observed statistic will lead to rejection of the null hypothesis than in the nondirectional test with the same rejection level.

Directional test. See **Directional hypothesis.**

Direct relationship. A relationship between two variables such that high values of one variable are found with high values of the second variable, and vice versa; the status of the relationship between two correlated variables is either positive or negative.

Discrete measurement. Measurement that can generate only certain values that are separated by discrete intervals.

Discrete variable. A variable that can assume only a finite number of values.

Distribution. The pattern of frequency of occurrence of scores; the total observations or a set of data for a variable; when observations are tabulated according to frequency for each possible score, we have a frequency distribution.

Distribution-free method. A method for testing a hypothesis or setting up a confidence interval, for example, that does not depend on the form of the underlying distribution.

Distribution-free tests. A term referring to a large family of statistical tests that, in general, do not require assumptions about the precise shape of the population distribution; the population distribution of a variable need not be normal in shape and data need not be at least interval level; also called **nonparametric tests**.

Dummy table. A cross-tabulation table that contains asterisks to reflect where disproportionately large frequencies will be found if a directional hypothesis is supported.

Dummy variable. A variable that is created by converting a qualitative variable into a binary variable.

Eigenvalue. In factor analysis, a quantity that corresponds to the equivalent number of variables that a derived factor represents.

Empirical frequency distribution. A frequency distribution tabulated from data that have actually been collected (as opposed to a theoretical frequency distribution, which is constructed from theoretical or mathematical considerations).

Empirical sampling distribution. A sampling distribution generated by actually taking random samples and measuring each sample's characteristics.

Error of estimation. Distance between an estimate and the true value of the parameter estimated.

Error of measurement. In measurement, the extent of its inaccuracy.

Estimate. Number computed from sample data used to approximate a population parameter.

Estimator. Rule that tells us how to compute an estimate based on data contained in a sample; an estimator is usually given as a mathematical formula, as in regression analysis.

Expected frequencies. In the chi-square test, the frequencies of observations in different categories (cells) that would be most likely to appear if the null hypothesis were true.

Expected value. The long-run average of a random variable over an indefinite number of samplings.

Expected value of a statistic. The mean of a statistic's sampling distribution.

Experiment. A research study in which we have control over the levels of the independent variable and over the assignment of people (objects) to different conditions.

Experimental group. In an experimental research design, the group in which the independent variable is manipulated or introduced.

External validity. The generalizability of a research finding to, for example, other populations, settings, treatment arrangements, and measurement arrangements.

Extraneous variable. See **Intervening variable.**

Factorial experiments. Experimental research designs that look at the separate effects and interactions of two or more independent variables at the same time.

Five-number summary. A concise description of the distribution of the values of a variable within a sample or population. It consists of the minimum value, the 25th percentile, the median, the 75th percentile, and the maximum value.

F ratio. The between-group estimates of the variance of the sampling distribution of the mean divided by the within-group's estimate; the F ratio is a measure of the strength of a treatment effect.

F statistic. A test statistic that is used to compare variances from two normal populations; used in analysis of variance.

Frequency. Number of observations falling in a cell or value category of a specific variable.

Frequency distribution. A table or graph that presents the number of times (frequency) with which different values of the variable occur in a group of observations; a technique of descriptive statistics that shows how often each score occurs.

Frequency polygon. A graphic technique of descriptive statistics that uses the height of connected dots to show how often each score occurs; graph of a frequency distribution in which the horizontal axis represents different values of a variable and the vertical axis represents frequencies with which those values occur; in constructing a frequency polygon, a dot is placed over each value of the variable at a height corresponding to the appropriate frequency; the dots are then connected with lines to form a polygon.

Frequency table. In its simplest form, a two-column table with one column listing values of a variable and the other column listing the frequency with which the different values occur within a group of observations.

Grouped cumulative frequency distribution. An extension of a grouped frequency distribution that shows how often scores occur at or below each interval.

Grouped frequency distribution. Table or graph in which frequencies are not listed for each possible value of the variable; rather, a frequency is listed for each of a number of intervals on the measurement scale; each interval is a range of values; all observations falling within the limits of the interval add to the frequency count for that interval; grouped frequency distributions are used most often when data represent observations on a continuous variable.

Grouped frequency histogram. A histogram that shows how often scores occur at given intervals.

Grouped frequency polygon. A frequency polygon that shows how often scores occur at given intervals.

Histogram. A graphic representation of a frequency distribution in which the horizontal line represents values of a variable and the vertical line represents frequencies with which those values occur; a bar is constructed over each value of the variable (or the midpoint of each interval, if the data are grouped) and extended to the appropriate frequency; the term *histogram* usually refers to such a graph for interval or ratio data, whereas the term *bar graph* usually refers to such a graph for nominal or ordinal data; a graphic technique of descriptive statistics that uses the heights of adjoining bars to show how often each score occurs.

Horizontal axis. The horizontal dimension of a two-dimensional graph; it usually represents values of the independent variable in frequency distributions; sometimes called the *x*-axis or the **abscissa**.

Hypothesis. See **Research hyopthesis.**

Hypothesis testing. A technique in inferential statistics in which we make a decision about the state of reality in the population; the decision consists of either accepting the state of reality proposed by the null hypothesis or rejecting the null hypothesis in favor of the research hypothesis; usually postulates a very specific set of conditions; a technique of inferential statistics that helps us decide whether research results are attributable to chance.

Hypothetical population. A statistical population that has no real existence but is imagined to be generated by repetitions of events of a certain type.

Independent groups *t* test. A statistical test used to decide whether two given independent samples could have occurred by chance.

Independent samples design. An experiment in which people are assigned to different groups on a completely random basis; samples are drawn in such a way that the particular subjects chosen for one sample have no influence on which subjects are chosen for the other sample.

Independent variable. The variable we believe to be associated with the different values of the dependent variable; the variable that is manipulated or introduced in a research study in order to see what effect differences in it will have on those variables proposed as being dependent on it.

Inferential statistics. Statistical methods that make it possible to draw tentative conclusions about the population based on observations of a sample selected from that population and, furthermore, to make a probability statement about those conclusions to aid in their evaluation.

Interaction. When the effect of one factor on a response depends on the level(s) of one (or more) other factor(s); the effect of one independent variable upon another; the failure of one independent variable to remain constant over the levels of another; two treatments are said to "interact" if scores obtained under levels of one treatment behave differently under different levels of the other treatment.

Interquartile range. A statistic used as a measure of variability; the distance between the 75th and 25th percentiles; the interquartile range is more stable than the simple range and can be used with ordinal level data; it does not, however, reflect the value of every observation in the group (as does the standard deviation); the median and interquartile range are often used together to describe a group, since both are based on percentiles.

Interval measurement. A measurement that, in addition to ordering scores, also establishes an equal unit so that distances between any two scores are of a known magnitude; a measurement in which objects, events, or processes are assigned to ordered categories that are separated by equal intervals; any measuring device that is capable not only of placing people (or objects) in their rank order on a characteristic but can also measure the differences between them in regard to that characteristic.

Intervening variable. A variable whose existence is inferred, but that cannot be manipulated; a variable that may affect just what influence (if any) an independent variable has upon a dependent variable; also referred to as a **confounding variable** or an **extraneous variable;** when controlled for in a research design, it is known as a **control variable.**

Inverse relationship. A relationship between two variables such that high values of one variable are found with low values of the other variable, and vice versa; sometimes referred to as a **negative relationship** or *negative correlation.*

Kurtosis. A quality of the distribution of a set of data dealing with whether or how much the data "pile up" around some central point; the quality of "peakedness" or "flatness" of the graphic representation of a statistical distribution.

Least-squares criterion. The principle that the best regression line is that one that would result in the smallest sum of squared deviations from the line.

Leptokurtic distribution. A relatively peaked frequency distribution; a frequency distribution that is more concentrated about the mean than the corresponding normal distribution.

Level of confidence. A term used in constructing confidence interval estimates of parameter values to specify our confidence that the interval includes the parameter value; using procedures for constructing a 95 percent confidence interval, for instance, we would enclose the true parameter value within its limits on 95 percent of such attempts; the higher the level of confidence, the wider the interval.

Level of measurement. Refers to the degree to which characteristics of the data may be modeled mathematically; the higher the level of measurement, the more statistical methods are applicable.

Level of significance. See **Rejection level.**

Limits of confidence intervals. The upper and lower values at the two ends of a confidence interval; in a symmetrical confidence interval, the limits are located one allowance factor above and below the sample statistic.

Linear correlation. A correlation between variables that if displayed using a scattergram would approximate a straight line.

Linear relationship. A relationship between two variables such that a straight line can be fitted satisfactorily to the points on the scattergram; the scatter of points will cluster elliptically around a straight line rather than around some type of curve.

Line of best fit. See **Regression line.**

Lower confidence limit. Smaller of the two numbers that form a confidence interval; in frequency distributions where data have been grouped, it is the lower boundary of an interval on the measurement scale.

Mann-Whitney *U* test. A nonparametric hypothesis-testing procedure used to decide whether two given independent samples would have arisen by chance from identically distributed populations; test for comparing two populations based on independent random samples from each.

Marginals. The count of frequencies with which certain responses occur; in a cross-tabulation table, the row and column totals.

Matched pairs test. A statistical test for the comparison of two population means; the test is based on paired observations, one from each of the two populations; in the two-sample experiment, a procedure in which the entire subject pool is arranged in matched pairs, where pair members are similar (matched) on important characteristics; one member of each pair is then assigned to each group.

Matrix. A two-dimensional organization; each dimension is composed of several positions or alternatives; any particular "score" is a combination of the two dimensions as, for example, in a correlation matrix.

Mean. A term shared by several measures of central tendency (arithmetic mean, harmonic mean, geometric mean, and quadratic mean), all of which are computed using the value of every observation in the data set; in a general sense, the mean is equivalent to the average of all of the values within a data set.

Mean deviation. Measure of variability that is literally the mean (absolute value) of the deviations about the mean.

Measurement. In the most general sense, the assignment of labels to observations according to a rule or system; in statistics, measurement systems are classified according to level of measurement and may produce data that can be represented in numerical form or in words; the assignment of numerals to objects or events according to specific rules.

Measure of central tendency. A single number that describes the location, or relative magnitude, of a typical score within a sample or population; synonymous with the term "average;" the mode, median, and mean are examples of central tendency.

Measure of variability. A single number that describes how spread out a group of scores is within a sample or population; the range, variance, and standard deviation are examples of variability.

Median. A measure of central tendency defined as the point on the measurement scale where 50 percent of the observations fall above it and 50 percent of the observations fall below it; it thus coincides with the 50th percentile; it is useful in skewed distributions because it is not as sensitive as the mean to the presence of a few outliers (extremely high or low values); it requires at least ordinal level data.

Mesokurtic distribution. A frequency distribution that is neither excessively peaked nor excessively flat; the normal distribution is a mesokurtic distribution.

Midpoint of an interval. The value located halfway between upper and lower limits of an interval, found by adding upper and lower limits and dividing by 2; when graphing or computing statistics from grouped data, the midpoint of each interval is sometimes used to represent all observations appearing in that interval.

Mode. A measure of central tendency; the most frequently occurring value in a distribution of scores (in grouped distributions, the midpoint of the interval with the highest frequency).

Most powerful test. The statistical test that has the smallest probability of producing a **Type II error.**

Multiple-group design. An experimental research design with one control group and several experimental groups.

Multivariate analysis. A statistical analysis of the relationship among three or more variables.

Mutually exclusive events. In applications of probability theory, two or more events that cannot both happen on a single trial; on a single flip of a coin, for example, the events "heads" and "tails" are mutually exclusive.

Negatively skewed distribution. See **Negative skew.**

Negative relationship. The situation in correlational analysis where high values of one variable tend to be associated with low values of another, and vice versa; negative relationships are indicated by negative correlation coefficients.

Negative skew. A descriptive term applied to frequency distributions with many high values and few extremely low values; on a frequency polygon, negative skew produces a "tail" in the direction of low values—to the left; skewness in which the mean is less than the mode.

Nominal measurement. A measurement that simply classifies elements into two or more mutually exclusive categories, indicating that elements are qualitatively different but not giving order or magnitude; a measurement in which objects, events, or processes are assigned to categories having no inherent order; the level of measurement whose only requirement is that each observation falls in one, and only one, measurement category; also referred to as categorical measurement; it is the lowest level of measurement.

Nondirectional test. A statistical test with two regions of rejection, that is, a two-tailed test; the area under the sampling distribution curve equal to the rejection level is divided into two equal parts at each end of the distribution, creating two regions of rejection; an observed statistic in either region leads to rejection of the null hypothesis; a test used when we have not predicted the direction of a relationship between two variables.

Nonparametric tests. Usually refers to statistical tests of hypotheses about population probability distributions, but not about specific parameters of the distributions; a test that does not require a normal population distribution; a method for testing a hypothesis that does not involve an explicit assertion concerning a parameter; hypothesis-testing procedures that do not make stringent assumptions about population parameters.

Normal distribution. A symmetrical, bell-shaped curve that often arises when a trait is composed of a large number of random, independent factors; the curve possesses a specific mathematical formula.

Null hypothesis. A statement concerning one or more parameters that is subjected to a statistical test; a statement that there is no relationship between the two variables of interest.

Observation. An objectively recorded fact or item of datum; statistics are usually applied to collections of observations.

One-sample *t* test. A hypothesis-testing procedure used to decide whether a given sample could have occurred by chance.

One-tailed test. See **Directional hypothesis.**

One-way analysis of variance. A statistical test used to decide whether two or more samples could have occurred by chance from populations with equal means.

One-way chi-square test. An application of the chi-square test in which observation categories differ only with respect to one variable (hence, one way); usually the test is concerned with the question of whether the observed frequencies are distributed in observation categories by chance or because of the influence of some variable other than chance.

Ordinal measurement. A measurement that classifies and ranks elements or scores; a procedure that is capable of rank ordering individuals (or objects) on a particular characteristic but that cannot distinguish how different each is from the others; a measurement in which objects, events, or processes are assigned to ordered categories; the level of measurement above nominal but below interval; the data represent at least ordinal scale measurement if each observation falls into one, and only one, category, and if observation categories can be rank ordered.

Ordinate. See **Vertical axis.**

Origin. The point of a graph at which the abscissa and ordinate intersect.

Outcome. A possible result of an experiment or observation; in probability applications, the result of an experimental trial; see also **Probability.**

Paired observations. An observation on two variables, where the intent is to examine the relationship between them; paired observations form the raw material of correlational analyses; recording both a person's height and weight and keeping both of those measurements associated with the same person constitute collection of a paired observation.

Parameter. A characteristic of a population determined from observations on every member of the population; population parameters of interest to us include the mean, range, median, standard deviation, and many others; also a characteristic of a mathematical relation whose value must be specified before the expression can be evaluated; a measure computed from all observations in a population.

Parameter estimates. Attempts to estimate the values of population parameters (e.g., the mean) from statistics computed on a sample selected from the population; estimates may consist of a single value (a point estimate) or a range of values (confidence interval).

Parametric tests. Statistical methods for estimating parameters or testing hypotheses about population parameters; a statistical test in which the null and research hypotheses are stated in terms of population parameter values; an example is the test for the significance of the difference between two means; procedures that make relatively stringent assumptions about population parameters.

Pearson's product-moment correlation coefficient. A correlation coefficient that specifies the strength and direction of a relation between two interval or ratio level variables; it is the most commonly used statistic in correlational analyses; also called *Pearson's* r.

Percent. Synonymous with "in 100" or the number of cases out of 100.

Percentage distribution. A table that displays the percentage of cases that were found to have each of the respective measurements of a variable.

Percentile. A point on the measurement scale below which a specified percentage of the group's observations fall; the 20th percentile, for instance, is the value that has 20 percent of the observations below it.

Percentile rank. A transformed score that tells us the percentage of scores failing at or below a given score.

Perfect relation. A relationship between two variables such that the value of one variable is known if the value of the other variable is specified; a relationship, either direct or inverse, in which there is a perfect predictability between the two variables; when all points in the scattergram lie exactly on the regression line.

Pie chart. A graph that displays the frequency distribution of a variable as portions of a circle reflecting percentages of the whole.

Platykurtic distribution. A frequency distribution that has a relatively flat shape; a distribution that is less concentrated about the mean than the corresponding normal distribution.

Point estimate. A single value, produced by application of inferential methods to observations on sample members, that is our best guess of a parameter value.

Population distribution. A distribution of all the scores in a population; a collection of all observations identifiable by a set of rules; a designated part of a universe from which a sample is drawn; the complete group of potential observations.

Positively skewed distribution. See **Positive skew.**

Positive relationship. The situation in correlational analyses that exists when high values of the first variable tend to be associated with high values of the second variable, and low values of the first variable tend to appear with low values of the second.

Positive skew. A descriptive term applied to frequency distributions with many low values and a few extremely high values; on a frequency polygon graph, positive skew produces a "tail" in the direction of the positive values; skewness in which the mode is less than the mean.

Power of test. See **Statistical power.**

Prediction. The estimation of scores on one variable from data about one or more other variables.

Predictor variable. The variable that, it is believed, allows us to improve our ability to predict values of the criterion variable.

Probability. A measure of likelihood; the number of outcomes in which an event can occur divided by the total number of possible outcomes; also called p.

Probability distribution. For discrete random variables, the probability distribution is a relative frequency distribution; the relative frequencies associated with values indicate the probabilities that they occur.

Proportion. A fraction of one.

Quartile. A percentile that is an even multiple of 25; the 25th percentile is the first quartile, the 50th percentile is the second quartile (it is also the median), and the 75th percentile is the third quartile.

Quartile deviation. A marking off of the distribution into four quarters so that we know the limits within which the middle half of the scores fall.

Quasi-experimental designs. Research designs in which research participants (or objects) cannot be assigned randomly to groups but where the independent variable can still be manipulated.

Random variable. A variable that can assume different values; there is a probability associated with occurrence of different values of the variable, and these probabilities constitute a probability distribution.

Range. Difference between the largest and smallest numbers of an array plus one; the distance between the highest and lowest values in a distribution (more accurately, the distance between the upper true limit of the highest value and the lower true limit of the lowest value); it is used as a measure of variability.

Ratio measurement. A measurement that, in addition to containing equal units, also establishes an absolute zero point within the scale; a measurement in which objects, events, or processes are assigned to ordered categories that are separated with equal intervals, and where the zero point is not arbitrary; the highest level of measurement; it is reached when each observation falls in one, and only one, category; when observation categories can be ordered; when there are equal intervals between adjacent categories on the measurement scale; and when a value of zero represents a zero quantity of the variable being measured.

Raw score. A numerical value assigned to an observation that is expressed in the original units of measurement; a score obtained directly by measuring some characteristic of a person, event, or process in a research study.

Regression analysis. A variation of a correlational analysis that makes possible prediction of the value of one variable from observations on another variable; these predictions are based on a collection of previously made paired observations on both variables; regression analyses require that the two variables be correlated and that the relation between them approximate a linear one.

Regression equation. A derived equation in the form of $Y' = a + b(X)$ that makes it possible to predict the value of a criterion value from a value of a predictor variable.

Regression line. A hypothetical line that goes through data points and that uses the method of least squares; one of two least-squares lines through a scatter plot of paired observations; each regression line constitutes the collection of predicted values for one of the variables; the straight line of best fit (usually according to the least-squares criterion) for a set of bivariate data; the line of best fit in a scatter plot; mostly used to predict values of the Y variable from values of the X variable.

Rejection level. Set of values of a statistical test that indicates rejection of the null hypothesis; a probability associated with the test of a hypothesis using statistical techniques that determine whether or not the null hypothesis is rejected; the commonly used rejection level is .05; probability of rejecting the null hypothesis when it is true; also called the *alpha level.*

Relative frequency distribution. A table or graph that shows observation categories and the proportion of the group that falls within each value category—that is, the relative frequency of each category; the proportion of observations that falls in one category or interval; in probability applications, the relative frequency of an event is the proportion of trials on which the event occurs.

Reliability. The consistency of a measurement instrument.

Reliability coefficient. A measure of the consistency of a statistical test; there are several methods of computing a reliability coefficient, depending upon the test and the specific situation.

Replication. Repetition of the same research procedures by a second researcher for the purpose of determining if earlier results can be duplicated; the collection of two or more observations under a set of identical experimental conditions.

Research hypothesis. A prediction that two or more variables will be found to be related; the hypothesis to be supported if the null hypothesis is rejected; also called the **alternative hypothesis.**

Rho. A population correlation coefficient.

Rival hypothesis. Theoretical alternatives for explaining the apparent relationship between the independent (or predictor) variable(s) and the dependent (or criterion) variable(s); other variables that might explain variations within the dependent (or criterion) variable(s).

Robustness. Refers to the property that certain hypothesis-testing procedures have of yielding the same decision regardless of whether all assumptions for the test are strictly satisfied.

Sample. A subset of the population under study; a subset of a population often used synonymously with "group" and "condition" when discussing research designs; sometimes referred to as a *research sample.*

Sample distribution. The frequency distribution of all observations in a sample; when a number of different samples are selected from one population, each sample will probably have a sample distribution slightly different from other samples.

Sample statistic. Characteristics of samples; statistics computed from observations on sample members; the mean of a sample is a sample statistic because only members of the sample contribute to its value; the mean of a population is a parameter (rather than a statistic) because all members of the population contributed to its value.

Sampling. A method of selecting members of the population for inclusion in a research study; strictly speaking, proper sampling procedures must be used if inferences about the population are to be made from sample statistics; two broad categories of sampling procedures are random (probability) sampling and nonrandom (nonprobability) sampling.

Sampling distribution. A theoretical distribution that can be specified for any statistic that can be computed for samples from a population; it is the frequency distribution of that statistic's values that would appear if all possible samples of a specified size N were drawn from the population; it is the foundation of inferential statistics because it allows one to specify the probability with which different values of the statistic appear; it is assumed that a statistic computed from sample observations is one value from such a distribution.

Sampling bias. Refers to the natural phenomenon whereby sample statistics tend to differ from population parameters; the degree to which the sample can be predicted to vary from the population in relation to some variable based on this phenomenon. Sometimes referred to as **chance.**

Sampling ratio. The ratio of sample size to population size; also referred to as *sampling fraction.*

Scattergram. A graphic representation of the relationship between two interval or ratio level variables; a two-dimensional graph in which each axis represents values of a different variable; paired observations on both variables are represented as dots on the graph; it may be used as a preliminary step in a correlational analysis or to portray in graphic fashion the strength and direction of a relationship between two variables; sometimes referred to as a *scatter plot.*

Score. A numerical value assigned to an observation; also called **data.**

Score interval. In a grouped frequency distribution, the range of observed values is divided into a number of score intervals; the frequency distribution table lists the number of observations that fall into each score interval.

Semi-interquartile range. Half the interquartile range, sometimes used as a measure of variability.

Sign test. Nonparametric statistical test used to compare the same sample at two different times.

Skewed distribution. A distribution in which more observations fall on one side of the mean than on the other side.

Skewness. A quality of the distribution of a set of data dealing with whether the data are (or are not) symmetrically distributed around a central point.

Spearman's correlation coefficient. A correlation coefficient showing the strength and direction of a relationship between ranks of two variables in a number of paired observations; it can thus be used when one or both variables produce data at the ordinal level; it is sometimes used as a quickly computed substitute for Pearson's *r*; also referred to as *Spearman's rho.*

Spurious relationship. Occurring by chance; not a "real" relationship.

Stability. The degree to which a statistic's value remains constant when it is computed for a number of different groups that are essentially alike but that differ in a few values; in inferential statistics, a stable statistic is one whose value remains stable from sample to sample, when all samples are taken from the same population.

Standard deviation. A common measure of variability; it requires at least interval level data and reflects the value of every observation in the distribution; like other measures of variability, it is a single number whose size indicates the spread, or dispersion, of the distribution; a measure of variability that is the square root of the **variance;** it represents a specified distance along the baseline of a distribution curve.

Standard error of a statistic. The standard deviation of the underlying (sampling) distribution of the statistic.

Standard error of estimate. In the regression situation, the standard deviation of observed values around the regression line; the smaller the standard error of estimate, the more precisely we can predict scores of the criterion variable.

Standard error of the difference. Refers to the standard deviation of the sampling distribution of the difference for independent samples.

Standard error of the mean. The standard deviation of the underlying (sampling) distribution of the mean.

Standard normal distribution. The normal distribution with a mean of zero and a standard deviation of one.

Standard score. A score stated in units of standard deviation from the mean of the distribution; a negative score indicates a score below the mean and a positive score indicates a score above the mean; an individual observation that belongs to a distribution with a mean of 0 and a standard deviation of 1; any distribution of raw scores can be transformed into a distribution of standard scores without changing the shape of the distribution or the relative order or distances between members because the transformation to standard scores is linear; also referred to as a *z* **score.**

Statistical decision. Choosing between states of possible reality on the basis of probability considerations; hypothesis testing involves a statistical decision in which we either accept or reject the null hypothesis.

Statistical power. The ability of a statistical test to reject correctly the null hypothesis; a test's ability to detect a true relationship between or among variables.

Statistically significant. Judged too unlikely to have occurred by chance. A statistically significant relationship between variables is, most likely, a real relationship.

Statistics. In comparison to the term *parameters,* statistics refers to the characteristics of a sample rather than to the characteristics of a population; in the context of descriptive statistics, measures taken on a distribution; in the context of inferential statistics, measures or characteristics of a sample; in a more general sense, the theory, procedures, and methods by which data are analyzed; the area of study that includes methods for producing and interpreting statistics; generally speaking, statistical methods are applied in an attempt to understand large masses of data, to discover and describe characteristics of the data that are not apparent from casual observation, and to describe characteristics of a group of observations rather than single observations.

Structural variables. In data analyses, those characteristics formed by combining units from lower levels of analysis.

Symmetrical distribution. A distribution in which, for every observation on one side of the mean, there is another observation at an equal distance on the other side of the mean; in a symmetrical distribution, the left half of the polygon (or histogram) is a mirror image of the right half; a distribution with a frequency polygon whose left and right sides will coincide if it is folded in the middle along a vertical line.

Transformed standardized score. A score that allows us to tell at a glance where it falls in a distribution of scores; a standard score that has been transformed so that it now belongs to a distribution with any mean and standard deviation we wish; transformed scores are used most often in evaluating test scores.

True limits of a number. The upper and lower points on the measurement scale that enclose all values of the variable actually represented by a number.

t **score.** A standard score that has been transformed to a distribution with a mean of 50 and a standard deviation of 10.

t **test.** A parametric hypothesis test that uses the *t* distribution to arrive at a decision; determines if there is a statistically significant difference between two means.

Two-tailed test. See **Nondirectional test.**

Type I error. An alpha error; error that occurs when the null hypothesis is rejected when it is true.

Type II error. A beta error; error that occurs when the null hypothesis is not rejected when it is false.

Unbiased estimator. Estimator that has a probability distribution with the mean equal to the estimated parameter; an estimate of a parameter is said to be unbiased if its expected value is equal to the parameter.

Unbiased statistic. A statistic computed in a manner such that the mean of its underlying distribution is the parameter that the statistic estimates.

Underlying sampling distribution of a statistic. The distribution (usually theoretical) of all possible values of the statistic from all possible samples of a given size selected from the population.

Unimodal. Refers to a distribution with only one mode.

Univariate analysis. Statistical analysis of the distribution of values of a single variable.

Upper confidence limit. Larger of the two numbers that form a confidence interval.

Upper limit of an interval. In frequency distributions where data have been grouped, the upper boundary of an interval on the measurement scale.

Validity. The degree to which a measurement instrument accurately measures what it is claimed to measure.

Variability. Dispersion of a distribution; the extent to which values differ among themselves; variability is not the name of a specific statistic; rather, it is the term applied to the characteristic of dispersion.

Variable. A characteristic that takes on different values; any attribute whose value, or level, can change; any characteristic (of a person, object, or situation) that can change value or kind from observation to observation.

Variance. Measure of variability that is the average value of the squares of the deviations from the mean of the scores in a distribution; measure of data variation; the mean squared deviation from the mean; the squared standard deviation, sometimes called the *mean square.*

Variation. Sum of the squared deviations about the mean; in some applications, variation is useful in its own right as a measure of variability; also called *sum of squares.*

Vertical axis. The vertical dimension of a two-dimensional graph; it usually represents frequency in frequency distributions, relative frequency in relative frequency distributions, and cumulative proportion in cumulative proportion graphs; when experimental results are graphed, it usually represents values of the dependent variable; also called the **ordinate.**

x **variable.** The variable plotted on the abscissa of a scatter plot, and the "predictor" variable (used to predict the *y* variable) in regression; usually the independent variable in a research study.

Yates's correction factor. In computing the obtained chi-square statistic, a mathematical correction that should be applied when $df = 1$.

y **variable.** The variable plotted on the vertical axis in a scatter plot, and the "predicted" variable (predicted from the *x* variable) in regression; usually the dependent variable in a research study.

Zero relationship. The situation that exists when values of one variable are not related in any way to values of another variable; with a zero relationship, knowing the value of one variable gives us no indication of the value of the other; perfect zero relationships are represented by correlation coefficients of 0.

z score. A transformed score that tells us how many standard deviations a score lies away from the mean in a distribution.

Z test. A hypothesis-testing procedure used to decide whether a given sample could have occurred by chance from a population with a given mean and known standard deviation.

References and Further Readings

Anderson, D. R., Sweeney, D. J., and Williams, T. A. (1986). *Statistics: Concepts and Applications*. St. Paul, MN: West.

Andrews, F. M., Klem, L., Davidson, T. N., O'Malley, P. M., and Rodgers, W. L. (1994). *A Guide for Selecting Statistical Techniques for Analyzing Social Science Data* (3rd ed.). Ann Arbor, MI: Institute for Social Research, The University of Michigan.

Blalock, H. M., Jr. (1979). *Social Statistics* (2nd ed.). New York: McGraw-Hill.

Bohrnstedt, G. W., and Knoke, D. (1988). *Statistics for Social Data Analysis* (2nd. ed.). Itasca, IL: F. E. Peacock.

Brown, R. W. (1992). *Graph It! How to Make, Read, and Interpret Graphs.* Englewood Cliffs, NJ: Prentice-Hall.

Craft, J. L. (1990). *Statistics and Data Analysis for Social Workers* (2nd ed.). Itasca, IL: F. E. Peacock.

Darlington, R. B., and Carlson, P. M. (1987). *Behavioral Statistics: Logic and Methods.* New York: Free Press.

Foddy, W. H. (1988). *Elementary Applied Statistics for the Social Sciences.* New York: Harper & Row.

Freund, J. E. (1988). *Modern Elementary Statistics* (7th ed.). Englewood Cliffs, NJ: Prentice-Hall.

Grinnell, R. M., Jr. (Ed.). (1993). *Social Work Research and Evaluation* (4th ed.). Itasca, IL: F. E. Peacock.

Guilford, J. P. (1950). *Fundamental Statistics in Psychology and Education* (2nd ed.). New York: McGraw-Hill.

Heyes, S. (1986). *Starting Statistics in Psychology and Education.* London: Weidenfeld and Nicolson.

Howell, D. C. (1987). *Statistical Methods for Psychology* (2nd ed.). Boston, MA: Duxbury Press.

Kachigan, S. K. (1991). *Multivariate Statistical Analysis* (2nd ed.). New York: Radius Press.

Khazanie, R. (1986). *Elementary Statistics: In a World of Applications* (2nd ed.). Glenview, IL: Scott, Foresman.

Kiess, H. O. (1989). *Statistical Concepts for the Behavioral Sciences.* Boston: Allyn and Bacon.

Krishef, C. H. (1987). *Fundamental Statistics for Human Services and Social Work.* Boston, MA: Duxbury Press.

Loether, H. J., and McTavish, D. G. (1988). *Descriptive and Inferential Statistics: An Introduction* (3rd ed.). Boston: Allyn and Bacon.

Miller, E. L. (1986). *Basic Statistics: A Conceptual Approach for Beginners.* Muncie, IL: Accelerated Development.

Reid, S. (1987). *Working with Statistics.* Cambridge, MA: Polity Press.

Shavelson, R. J. (1988). *Statistical Reasoning for the Behavioral Sciences* (2nd ed.). Boston: Allyn and Bacon.

Stahl, S. M., and Hennes, J. D. (1980). *Reading and Understanding Applied Statistics* (2nd ed.). St. Louis: Mosby.

Weinbach, R. W., and Grinnell, R. M. Jr. (1995). *Applying Research Knowledge: A Workbook for Social Worker Students.* Boston: Allyn and Bacon.

Wilcox, R. R. (1987). *New Statistical Procedures for the Social Sciences: Modern Solutions to Basic Problems.* Hillsdale, NJ: Lawrence Erlbaum.

Williams, M., Tutty, L. M., and Grinnell, R. M., Jr. (1995). *Research in Social Work: An Introduction* (2nd ed.). Itasca, IL: F. E. Peacock.

Wright, S. E. (1986). *Social Science Statistics.* Boston: Allyn and Bacon.

Yegidis, B., and Weinbach, R. W. (1996). *Research Methods for Social Workers* (2nd ed.). Boston: Allyn and Bacon.

Index